EXCLUSION

EXCLUSION

Homosexuals and the Right to Serve

———

MELISSA WELLS-PETRY
MAJOR, U.S. ARMY

Preface by Admiral Thomas H. Moorer,
U. S. Navy (Ret.)

REGNERY GATEWAY
Washington, D. C.

The views expressed in this book are those of the author and do not reflect the official policy or position of the Department of Defense or the U.S. Government.

Library of Congress Cataloging-in-Publication Data

Wells-Petry, Melissa, 1957–
Exclusion : homosexuals and the right to serve / Melissa Wells-Petry.
p.   cm.
Includes bibliographical references and index.
ISBN 0-89526-504-4 (acid-free)
1. United States—Armed Forces—Gays. I. Title.
UB418.G38W45   1993
355'.008664—dc20   93-21754
                                               CIP

Published in the United States by
Regnery Gateway
1130 17th Street, NW
Washington, D.C. 20036

Distributed to the trade by
National Book Network
4720-A Boston Way
Lanham, MD 20706

Printed on recycled, acid-free paper.

Manufactured in the United States of America.

10   9   8   7   6   5   4   3   2   1

To all the magnificent, muddy, real soldiers
with whom I have served—

"These Are My Credentials"

I could have never seen this project through without the constant encouragement of some of the best "soldiers" in the world. To friends and family, and especially to my husband, I give my warmest thanks. Thanks, too, to Professor William Woodruff for his wise advice and support. My applause and compliments to each person in the publication process who worked so hard to bring this book to life.

Contents

Preface

President Clinton's intention to lift the ban on homosexuals in the military has delivered our armed forces into the hands of an unnecessary fire fight. Not since Vietnam has the military been the focus of so much negative publicity. Like Vietnam, the energy level of the debate indicates the potency of the political forces at work.

Homosexuals have chosen this issue as a landmark of success. Aided by the powerful allies of the media and the present administration, homosexual activists have redirected the debate. When they speak misty-eyed of being denied their constitutional rights, some equate their movement—incorrectly—with the civil rights movement. Anyone who disagrees, particularly anyone with moral objections, is accused of being backward, ignorant, and discriminatory. But it is not—neither logically, nor legally—a question of civil rights. The matter of homosexuals in the military is a matter of national defense. In this duel over the future of America's ability to fight wars, the homosexual rights movement has chosen the weapons, picked the site, and taken advantage of a commander in chief with no military experience.

Exclusion evens the odds. Major Melissa Wells-Petry examines all the arguments from a legal and constitutional perspective. She is the first to discuss the military's point of view without rancor, without ulterior motives, and without media interruption and distortion. *Exclusion* is a true understanding of the issue.

The crux of the military's argument is the purpose of our military. That purpose—and the risking of lives, the training of recruits, the spending of taxpayers' money—is to defend our country. Our armed forces are not a paradigm of American democracy, and they are not meant to "look like America." There is one purpose, and if we no longer use the military for that purpose, then we no longer need a military. To suggest that we use the military to balance social inequalities displays a profound and disturbing ignorance of its mission. This demand, as Major Wells-Petry demonstrates, is also legally invalid.

Twenty years and dozens of legal cases against the armed services have produced a virtually unanimous body of law affirming the right of the military to determine their standards of service. There is no such thing as a constitutional right to serve in the military, just as there is no such thing as a constitutional right *not* to serve in the military. As Major Wells-Petry points out, to serve is both a privilege and a burden, but it is not a right. People may be prohibited from serving in the military for a variety of reasons, such as height, weight, religious persuasion, political views, educational record, and their sexual orientation.

Exclusion policies exist to select groups of people whose physical, emotional, and mental qualities give them the best potential for combat. These policies make our military efficient and effective. Some have pointed out that homosexuals have served our country bravely and well and claimed this as proof that a homosexual is capable of being a great soldier. No doubt this is true, but the argument is extraneous. Admission policies by their natures choose *groups* of people, not individuals. In determining its admission policy, the military must consider homosexuals as a group.

Major Wells-Petry discusses in some detail the military's objections to lifting the ban. One repercussion of allowing admitted homosexuals in the military would be the effect on privacy. Nowhere are men and women forced to shower together, to sleep together, and to receive mass physical exams together. Gender segregation, she explains, is based on two principles: that people have a sexual preference for people of the opposite sex, and that they should be allowed to choose to whom they expose an aspect of their sexuality. We should not challenge such well-established assumptions simply to accommodate a relative handful of homosexual activists who have presented no argument that to do so would make the military more effective.

When we consider changing military policy, the first question we should ask is "Will it help us win wars?" If the answer is "no," or even "we do not know," then we should make no such change. In the case of lifting the ban on homosexuals in the military, we have a simple choice: we can further the interests of a partisan political group, or we can strongly oppose any action which will degrade the combat readiness of our men and women in uniform.

Admiral Thomas H. Moorer,
U. S. Navy (Ret.)

EXCLUSION

CHAPTER ONE

Introduction

Others will debate the controversial issues . . . which divide men's minds; but serene, calm, aloof, you stand as the nation's war-guardian . . . as its gladiator in the arena of battle. . . . Your guidepost stands: Duty— Honor—Country.

> DOUGLAS MACARTHUR
> General of the Army
> West Point, 1962[1]

Serene, calm, aloof, the nation's war-guardian. For many, General MacArthur's famous address to the cadets at West Point set forever the standard for military service—the touchstone for understanding the military institution. Yet over the three decades since Mac-Arthur spoke the principles of "duty, honor, country," controversies surrounding the United States Army have rarely been phrased in such noble terms.

Indeed, today we hear dissonant voices proposing other definitions of military service, other roles for the nation's war-guardians— roles phrased in terms of individual rights, social goals, and political agendas. Our armed forces, we are told, should take the lead in social experimentation and adopt as a chief objective the righting of what some groups see as ancient societal wrongs. In the wake of this press for a new attitude toward the role of the military, President Clinton is considering a recommendation that he issue an executive order ending the Defense Department's long-standing ban on homosexuals in the services.

The debate over this question is already intense. Indeed, thus far the issue seems to have generated more heat than light, emotion without reasoned analysis. So in an attempt to clarify an important and technically complex subject for a general audience, this book will examine the controversy occasioned by the military's homosexual exclusion policy.[2]

In order to focus more sharply on the specific questions raised in the current debate, the book will concentrate on the U.S. Army as an exemplum of how the policy operates for all military branches, as well as of what issues all branches of military service will face if the present policy is altered. While a few arguments and examples may be more pertinent to the Army than to other branches, the book as a whole will deal with problems common to all the armed forces.

For those not familiar with military operations and terminology, some clarification is in order. The term "exclusion" is used here and elsewhere to mean disqualification from military service in general. Thus, an individual may be "excluded" from military service at recruitment, at reenlistment, or at any time administrative or judicial authorities take action to separate the individual from the military on the basis of homosexuality or some other service disqualifier.

Homosexuals are legally excluded from military service based on the "secretarial determination" that homosexuality is incompatible with military service.[3] The term "secretarial determination" refers here to the long-standing authority of the service secretaries to determine who is and who is not qualified for military duty. The controversy over whether or not the Secretary of Defense's determination that homosexuality is incompatible with military service has been phrased in constitutional terms, in terms of fairness, and even in political and sociological terms.

Putting that controversy in context, however, requires phrasing it in *military* terms. Those grappling with the issue must recognize that the debate is really over how to compose a *military fighting force whose mission is to be the nation's "gladiator in the arena of battle."* Only then can the American public fully understand the dynamic of the homosexual exclusion policy.

Far from heeding General MacArthur's advice that the Army should stand aloof from national or social controversies—in this instance, controversies regarding homosexual rights—those who oppose the homosexual exclusion policy have attempted to draw the Army directly into the fray. Homosexuals charge that the military "discriminates." Ugly, suggestive of ultimate evil, the word "discrimination" has come to mean the very antithesis of our finest American ideals. To label a policy "discriminatory" seems automatically and absolutely to judge it as morally wrong and legally indefensible. Yet, clearly *all military personnel policies discriminate.*

The military in fact discriminates on a variety of bases. For example, the military excludes—"discriminates against"—single parents,[4] felons,[5] handicapped individuals,[6] transsexuals,[7] conscientious objectors,[8] and persons with any of a number of medical conditions.[9] The military also discriminates on the basis of height and weight,[10] physical and mental ability,[11] visual acuity,[12] political beliefs and religious affiliation,[13] language,[14] youth and age.[15] To repeat—*all military personnel policies discriminate.* They discriminate between individuals or groups that have strong potential for successful soldiering and those that do not. And these discriminatory judgments are made by Congress, by the Secretary of Defense, or by the service secretaries in fulfilling their duty to compose strong, combat-ready, and efficiently administered armed forces.

Implicit in military personnel policies is the well-recognized legal principle that *no one has a right to military service.*[16] In the same way, *no one has the right to avoid military service* when called upon by the nation to serve.[17] This twofold principle—that there exists neither a right to serve nor a right to avoid service—flows from the military's responsibility to compose the nation's fighting forces based on the needs of the service. When the Army is viewed as what it is —the nation's war-guardian—little room exists to doubt the wisdom of such a principle.

Military service at best is an inestimable privilege; at worst it is a harsh and dangerous burden. This unique truth must illuminate and inform any debate over the creation of a "right" to military service for homosexuals—or for any other group.

The discussion that follows will be guided by three broad themes: the secretarial prerogative to establish the homosexual exclusion policy; the courts' traditional disposition to leave policy making to the secretaries; and the authority of Congress to change the policy through the democratic process. In addition, the discussion will provide an overview of the legal challenges to the military's homosexual exclusion policy, as well as a more detailed examination of the regulatory rationale for the homosexual exclusion policy and its practical implications. Finally, several proposals to modify the exclusion policy will be reviewed.

CHAPTER TWO

THE SOUND AND THE FURY:

The Temptation to Equate Controversy with Substance

OVERVIEW OF THE LITIGATION, 1971–1991

The sound and the fury surrounding recent legal challenges to the military's exclusion of homosexuals has created the impression that there is widespread discontent with the policy. Moreover, some have cited the intensity of the debate as evidence of a great ground swell of support to change the policy and make it more favorable for homosexuals. Whether this sound and fury illuminates or obscures the practical issues surrounding homosexuality within the military is a question discussed in a subsequent section, one that addresses the rationale for the current policy and the politics of social experimentation. However, after considering the legal history of the question, it is easy to see that the din accompanying current challenges to the homosexual exclusion policy is far out of proportion to the facts.

More than ten million people have served in the armed forces since 1971. In the last twenty years, the United States Army has defended thousands of lawsuits based on various kinds of military personnel policies. Many of these lawsuits were brought by individuals who sought to get into the Army or to get out of the Army—and on every conceivable ground.

Over the last twenty years, six cases brought by soldiers challenging the homosexual exclusion policy were decided.[1] Also, over these twenty years, six cases challenging the homosexual exclusion policies of the other military branches were decided.[2] Of these twelve lawsuits, none was filed by an individual who was denied enlistment, commissioning, or admission to a commissioning program by reason of homosexuality. Rather, the cases generally were filed by individuals who entered the military knowing homosexuality was a service-disqualifier, and who then challenged the exclusion policy once their homosexuality was discovered.

Some homosexuals have described how they entered the military by evading enlistment-interview questions about their homosexuality or by deceit.[3] One former service member stated:

I knew that being gay in the military was taboo. . . . I think every person in . . . the service has filled out the same form. They ask you . . . "Are you now or have you ever been a homosexual?" I'd been to bed with [other] men . . . but when I saw that question, there was no hesitation—I lied. I nevertheless wanted to participate in sexual activity with men, wanted to meet other men that wanted the same thing, and thought I could keep my activities under wraps.[4]

Another former service member described how he "made it through" the enlistment interview:

I knew that the regulation against homosexuality could keep me out of the military. . . . I do remember very clearly how the military doctor stated the question about homosexuality, "Do you have any problems with homosexuality . . . ?" he asked. "No," I answered. . . . I remember . . . coming home to my partner and saying, "I made it through. I didn't have to lie." And as far as I was concerned, I did not lie. . . . I had a right to be there, and I felt that what I was doing was not against the regulations.[5]

In a third example, a former service member explained, "I certainly side-stepped that god-awful question about homosexuality. I said, 'no.' I lied. I knew I was telling an untruth, but who was to know?"[6] In Dubbs v. Central Intelligence Agency, an example from civilian life, a plaintiff challenging the CIA's policy on homosexuality had failed to disclose her sexual orientation in the initial interview.[7]

Evasion or deception during the interview process is not unique to homosexuals. In Dillard v. Brown, for example, a case challenging the single parent exclusion policy, plaintiff had not disclosed to the military that she had a minor child.[8] Moreover, whether by chance, design, or error, a few homosexuals were enlisted in spite of the fact they acknowledged homosexuality. In Woodward v. United States, plaintiff, who enlisted in 1972, at that time acknowledged sexual attraction to other men, but denied ever engaging in homosexual conduct.[9] In Watkins v. U.S. Army, plaintiff was drafted in 1967 after an Army psychiatrist discounted his statement that he was homosexual.[10]

Watkins probably received more publicity than any other case challenging the homosexual exclusion policy. On first hearing, one point in his favor was the fact that he had informed an Army doctor of his homosexuality. From this, it was one small leap for a court unfamiliar with the military to conclude the Army "knew" Watkins was homosexual, assumed the risk in enlisting him, and was therefore unfairly discharging him.[11] This leap is not so small, however, when one considers the historical context of Watkins' enlistment. One former soldier explained:

> [The military] had a big problem with the gay issue around this same time because of Vietnam. *There were an awful lot of guys who'd march in and say, "Hey, I'm gay—send me home."* So the military had a very bizarre policy for a while: that if you came in and said you were gay, you had to be able to prove it . . . because of people trying to use homosexuality to evade the service.[12]

In 1968, a Marine was convicted of falsely swearing that he and other Marines were homosexual in order to obtain a discharge from the military.[13]

With these two exceptions, *Woodward* and *Watkins*, the cases challenging the homosexual exclusion policy were filed by individuals who knew their homosexuality was a service-disqualifier, entered the service anyway, and then brought suit when the Army enforced the policy. Of the twelve cases challenging the homosexual exclusion policy, no court finally held the policy legally infirm on any ground.[14] Thus, despite the contrary impression created by recent publicity and by some academic commentators, a substantial and virtually unanimous body of law affirms the constitutionality of the homosexual exclusion policy. This chapter will review this body of law as it pertains to privacy rights and First Amendment freedoms. The next chapter will review the law pertaining to due process and equal protection guarantees.

HOMOSEXUALITY AND THE RIGHT TO PRIVACY

One common theme in legal challenges to the homosexual exclusion policy has been "infringement of the right to privacy." Similar privacy issues have been raised and rejected in challenges to other military personnel policies[15] and statutory schemes.[16]

Privacy, as a constitutional right, primarily "is rooted in the First, Fourth, Fifth, and Ninth Amendments."[17] The Supreme Court has delineated the scope of the right to privacy by decision, rather than by definition. The Court exemplified this "decisional" (rather than "definitional") approach by recalling a little used word—"penumbra"—to active duty. The penumbra concept is the analytical equivalent of drawing circles around explicit constitutional rights and declaring the right to privacy is in *there*—in the "penumbra," which literally means the partial illumination between the dark shadow and the full light.

The penumbra, as one court observed, "was no more than a perception that it is sometimes necessary to protect actions or associations not guaranteed by the Constitution in order to protect an activity that is."[18] Thus, privacy cases have recognized rights that have little or no textual support in the language of the Constitution.[19] In challenges to the homosexual exclusion policy, an additional privacy question must be raised: Is there *any* right to privacy in a military environment?[20]

The Supreme Court has held that a constitutional right to privacy may not be substantially abridged by government regulation of matters regarding family relationships,[21] marriage,[22] and child-bearing.[23] The Court repeatedly has declined to expand the right to privacy beyond these areas.[24] In Bowers v. Hardwick, the Court explained its reluctance to forge ever-new boundaries for penumbral rights:

> Nor are we inclined to take a more expansive view of our authority to discover new fundamental rights imbedded in the Due Process Clause. The Court is most vulnerable and comes nearest to illegitimacy when it deals with judge-made constitutional law having little or no cognizable roots in the language or design of the Constitution. . . . There should be, therefore, great resistance to expand the substantive reach of those Clauses, particularly if it requires redefining the category of rights deemed to be fundamental.[25]

Further, the Court frequently has held that, even in these sensitive and intimate areas, some regulation is permissible. In Bradbury v. Wainwright, for example, the court noted, "the right to marry is not unfettered. . . . In addition to regulating the procedures, duties, and rights stemming from marriage, state regulations have absolutely prohibited certain marriages, such as result by incest, bigamy, or homosexuality."[26]

Plaintiffs challenging the homosexual exclusion policy on privacy grounds have cited two theories. The first theory asserts a constitutional right to privacy in homosexual conduct.[27] The second theory asserts a constitutional right to privacy in simply being homosexual.[28] Both theories have failed to stand up under close legal scrutiny. Although courts deciding challenges to the homosexual exclusion policy have vigorously debated the scope of the right to privacy,[29] no court has extended constitutional protection to homosexual conduct.

One court held "even if privacy interests were implicated [by the homosexual exclusion policy], they are outweighed by the Government's interest in preventing armed forces members from engaging in homosexual conduct."[30] Courts have recognized that other government interests, such as public health concerns, also may outweigh privacy interests.[31]

The issue of privacy in homosexual acts finally was settled in

Hardwick. There the Supreme Court rejected claims the "right" to engage in sodomy was "fundamental" and therefore protected by the Constitution.[32]

The second privacy theory focused, not on conduct, but on the state of being homosexual—a "psycho-sociological state of being" defined variously as the sum of the "integral components of one's personality";[33] an expression of "fundamental matters at the core of one's personality, self image, and sexual identity";[34] "the essence of one's identity";[35] "homosexual status";[36] and one's "sexual preferences."[37] Plaintiffs urged that the homosexual exclusion policy burdened this psycho-sociological state of being, and thus abridged the constitutional right to privacy.

In general (and not only in regard to the homosexual exclusion policy), the heart of any argument based on a burden on a "psycho-sociological state of being" is the notion the law has the power to control a person's *thoughts* or *feelings*, rather than the power to control only conduct. This notion is often uncritically dismissed as a truism. Nevertheless, every individual instinctively understands that this proposition is untrue. Indeed, that is the dilemma for the individual: the law proscribes certain conduct, yet thoughts and feelings associated with that conduct remain as strong as ever. The fact is, "freedom to think is absolute *of its own nature*, [but] the right to express thoughts . . . at any time or place is not."[38] Thus, in challenges to the homosexual exclusion policy, the privacy argument based on a burden on some "psycho-sociological state of being" cannot survive logical or legal analysis.

Indeed, courts did not find that the homosexual exclusion policy burdened any right to privacy, whether based on homosexual acts or on a state of being.[39] Just as denial of the opportunity to be a soldier does "not affirmatively curtail marriage or child bearing" for single parents,[40] courts found that the denial of the opportunity to be a soldier does not infringe on any asserted right to be homosexual.

In sum, the constitutional right to privacy does not protect homosexual conduct, nor does it insulate homosexuals from the impact of otherwise permissible regulations. The homosexual exclusion policy, therefore, is not invalid on any theory of a constitutional right to privacy.

HOMOSEXUALITY AND FIRST AMENDMENT RIGHTS

In addition to privacy claims, plaintiffs argued that the homosexual exclusion policy abridged constitutional rights of free association and free speech. These issues had been heard before in cases challenging other military exclusion policies, and they were resolved similarly.

THE RIGHT TO FREE ASSOCIATION

The right of association, like the right to privacy, is not express, but rather emanates from the First Amendment.[41] The Supreme Court recently affirmed that the right to associate is protected constitutionally in two distinct situations—"intimate association" and "expressive association."[42]

The phrase "intimate association" pertains to the association in human relationships that are fundamental to personal liberty. The phrase "expressive association" pertains to the exercise of rights expressly guaranteed by the First Amendment—such as speech, assembly, and petition for redress of grievance. In Dallas v. Stanglin, the Court again recognized practical limits on associational rights, stating, "[i]t is possible to find some kernel of expression in almost every activity a person undertakes—for example, walking down the street, or meeting one's friends at a shopping mall—but such a kernel is not sufficient to bring the activity within the protection of the First Amendment."[43]

Regardless of what type of association is at issue, the underlying activity must be protected by the First Amendment before constitutional guarantees apply.[44] In other words, if homosexual conduct is not protected by the Constitution, then there is no right to associate for that purpose. In these homosexual cases, plaintiffs' theories began with rights emanating from the First Amendment—privacy and associational rights—and then asserted that these rights operated to protect homosexual acts. Thus, by stating that homosexual conduct was protected *because* the right of association was protected, plaintiffs argued the converse of the established principle that association is protected only when the underlying conduct is protected.

In challenges based on associational rights, plaintiffs raised issues of both intimate and expressive association. In Berg v. Claytor[45] and Dronenburg v. Zech,[46] for example, plaintiffs claimed, in essence, that the association incident to homosexual acts was "intimate association" for purposes of First Amendment guarantees.[47] Neither court addressed the question of whether or not homosexual conduct included a protected sphere of intimate association. Courts simply found exclusion from military service "based on homosexual activity 'places no restriction on [an individual's] right to associate with whomever he chooses, and clearly does not contravene the First Amendment.' "[48] This holding shows how courts correctly began their reasoning with the underlying conduct, not with the rights asserted to emanate from the conduct.

The concept of "expressive association" was at issue in cases in which the record lacked evidence of homosexual acts. The failure of the record to establish homosexual acts is relevant to resolution of the legal issues because the court must decide *on the basis of the record*. The fact that no homosexual acts are established on the record, however, does not necessarily or logically indicate that the plaintiff has not committed homosexual acts.

The entirely appropriate rule confining the court to facts in evidence, thus, still raises issues of permissible and appropriate inferences to be drawn from facts in evidence. In Ben-Shalom v. Marsh, for example, the Army did not establish acts, but plaintiff expressly declined to deny "she has engaged or will engage in homosexual conduct."[49] In *Woodward*, plaintiff did not claim celibacy, but the court rejected counsel's attempts to characterize plaintiff as having only homosexual "tendencies".[50] In Pruitt v. Weinberger, the court did not make findings on whether or not the Army had established homosexual acts, but the record showed plaintiff twice was "married" to another woman and marriage presumably embraces some type of sexual consummation.[51]

Besides the issue of drawing reasonable inferences from facts in evidence, the sexual nature of the conduct in question often posed practical and legal obstacles to getting facts on the record. In Steffan

v. Cheney, the Navy did not establish homosexual acts on the record, but plaintiff obtained a restraining order to avoid answering questions about his commission of homosexual acts. Moreover, plaintiff did not challenge the Navy's right to refuse reinstatement based on commission of homosexual acts.[52] In a similar instance in *Aumiller* v. University of Delaware, the university did not establish that the professor engaged in homosexual acts, but he obtained a restraining order to prevent the university from inquiring into his living arrangements with a homosexual student.[53] Though the facts of record are silent, plaintiffs Steffan and Woodward disclosed their homosexual acts in interviews.[54]

In sum, the stilted nature of the facts of record in these cases frequently requires analytical and procedural gymnastics. As explained in *Cyr v. Walls*, "the Court can take notice of the logical distinction between gay individuals who simply prefer the companionship of members of their own sex and homosexual individuals who actively practice homosexual conduct. . . . This entirely theoretical distinction could, of course, be redetermined on the basis of later factual proof."[55]

Ben-Shalom v. Marsh was the only case explicitly to address expressive associational rights.[56] Associational rights initially were raised in *Matthews v. Marsh*[57] and *Woodward*.[58] The claim was dropped in *Matthews* because the plaintiff later admitted to homosexual acts. In *Woodward*, the claim was not pursued on appeal.[59]

In *Ben-Shalom*, the district court read the Army's then-current regulation as infringing a soldier's right "to meet with homosexuals and discuss current problems or advocate change in the status quo . . . [or] to receive information and ideas about homosexuality."[60] This aspect of the court's reasoning comports with settled analysis of speech issues, that "[the] critical line for First Amendment purposes must be drawn between advocacy, which is entitled to full protection, and action, which is not."[61]

The court's reading was based on regulatory language mandating discharge of a soldier who "evidences homosexual tendencies, desire, or interest."[62] This regulatory language subsequently was clarified. Thus, it is inaccurate to state, as some commentators have, that the Army's *"current* [homosexual exclusion] policy allows separation [from the military] based on homosexual tendencies alone."[63] As the court in *Ben-Shalom* clearly explained:

> [T]he new regulation we are now considering is a different regulation from that originally considered in *Ben-Shalom I*. The features then found objectionable have been substantially eliminated. Discharge of a soldier who "evidences homosexual tendencies, desire, or interest" is no longer broadly mandated. . . . The "tendency" and "interest" language is now gone. . . . What remains in the new regulation is but a bar against persons who either admit a "desire" to commit homosexual acts or who have in fact committed homosexual acts. . . . We are here concerned with plaintiff's forthright admission that she is a homosexual. That reasonably implies, at the very least, a "desire" to commit homosexual acts.[64]

The difference between "tendencies" and "desires" is critical and substantial. A person may tend to gain weight, for example, even though he does not desire to gain weight. Thus, desire has a volitional aspect while a tendency may be involuntary.

On review, the Circuit Court of Appeals in *Ben-Shalom* found that association—expressive or otherwise—was not implicated, much less chilled, by the homosexual exclusion policy.[65] Indeed, in cases in which a threat of criminal prosecution was alleged to chill free association—and certainly prosecution is more serious than an administrative sanction—the Supreme Court has stated even if the threat of prosecution had a deterrent effect on association,

> [T]he restraint is at most an indirect one resulting from self-censorship, comparable in many ways to restraint resulting from criminal libel laws. The hazard of such restraint is too remote to require striking down a statute which on its face is otherwise plainly within the area of congressional powers and is designed to safeguard a vital national interest.[66]

The homosexual exclusion policy is "plainly within the area of congressional powers and is designed to safeguard a vital national interest."[67] The homosexual exclusion policy, like other exclusion policies, simply denies individuals the opportunity to be soldiers.

The policy does not deny or preclude homosexuals the opportunity to engage in constitutionally protected intimate or expressive association with others.

THE RIGHT TO FREE SPEECH

Although privacy and associational claims were advanced, freedom of speech was the principal focus of First Amendment challenges to the homosexual exclusion policy.[68] Analysis of First Amendment speech questions is well settled. The first inquiry is whether or not the factual predicate of the claim constitutes speech. In United States v. O'Brien, for example, the Supreme Court noted "[the law] prohibits knowing destruction of certificates issued by the Selective Service System, and there is *nothing necessarily expressive about such conduct.*"[69]

If speech is at issue, the court must inquire whether or not that speech is constitutionally protected.[70] If protected speech is at issue, the next inquiry is whether the regulation actually implicates or abridges that speech. In *O'Brien*, the Court demonstrated this inquiry by holding that "[a] law prohibiting destruction of Selective Service certificates no more abridges free speech on its face than a motor vehicle law prohibiting the destruction of drivers' licenses."[71]

Finally, if it affects protected speech, the government's interests in the regulation are evaluated. The regulation must be within the constitutional power of the government, further an important or substantial government interest, and not restrict First Amendment rights more broadly than required to further the government's interests.[72]

The declaration of homosexuality as speech

Most challenges to the homosexual exclusion policy based on the First Amendment right to freedom of speech involved the plaintiff's declaration, "I am a homosexual." Courts found the statement, "I am a homosexual," to be evidence of identity or admission of a fact about oneself, rather than speech per se.[73]

Although the statement "I am a homosexual" is, "in some sense speech, it is also an act of identification. And it is the identity that makes [one] ineligible for military service, not the speaking of it aloud."[74] Likewise in Johnson v. Orr, the court viewed "plaintiff's self-assertion of her homosexuality as nothing more than an admission of a fact, and such fact may serve as a lawful basis for discharge."[75] Simply put, it is impossible to "accept the view that an apparently limitless variety of conduct can be labeled 'speech' whenever the person engaging in that conduct intends thereby to express an idea."[76]

Statements of identity frequently disqualify individuals for military service without the slightest constitutional moment. Statements such as "I am a single parent," "I am overweight," and "I am not a high school graduate," for example, trigger the application of military personnel regulations and may result in exclusion.[77] The statement, "I am a Nazi," certainly is protected political speech, quite unlike a statement of one's sexual identity. Still, in Blameuser v. Andrews, the Court of Appeals readily upheld excluding an individual from the military who—by his statement—identified himself as a Nazi.[78] No different result is warranted for statements of sexual identity, marital status, religious affiliation, health or educational status, or for any other statement that is "simply an admission that [the individual] comes within a classification of people whose presence in the Army is deemed by the Army to be incompatible with its . . . goals."[79]

As the Court held in Ben-Shalom, the operation of all military personnel exclusion policies plainly is triggered by the identity that disqualifies the individual for military service. Under these policies, how the disqualifying identity is revealed is simply immaterial. Thus, as one court frankly concluded, "the free speech issue raised in . . . [cases challenging the homosexual exclusion policy] appears to be specious."[80]

The declaration of homosexuality as protected speech

Even if the statement, "I am a homosexual," constitutes speech, further inquiry is required to determine if that speech is protected by the First Amendment. For government employees—for these purposes, a group similarly situated with soldiers—the standard-bearer for protected speech is Connick v. Myers.[81] *Connick* held for speech to be within the ambit of the First Amendment, it must address a "matter of public concern."[82] An individual's declaration of homosexuality does not meet this standard.

Looking first to the content of the statement, courts held declarations of one's homosexuality involve "facts private in nature."[83] Clearly, the mere event of publication is not sufficient to bring a declaration of one's homosexuality within the First Amendment.[84] Moreover, such declarations generally are made "for personal reasons and not to inform the public of matters of general concern."[85]

The opposite conclusion—that is, that declarations of homosexuality *are* made for the public's benefit—was advanced, accepted by some judges, but ultimately rejected. In *Matthews*, the lower court's analysis was that "Matthews' self-identification [as a homosexual] was undoubtedly a matter of significant 'personal interest' to her. Placing it in the latter category, however, ignores the reality that the issue of open employment for homosexuals (particularly in military service) is currently a matter of intense public debate."[86] In Rowland v. Mad River Local School District, the dissent argued an individual's declaration of homosexuality "necessarily and ineluctably" involves that individual in "public debate . . . regarding the rights of homosexuals."[87]

These arguments have confused public statements with public debate. To be part of public debate, statements must have an aspect of advocacy. Operation Desert Storm, for example, engendered debate on deployment of single parents to combat zones. But an individual's declaration that he is a single parent does not necessarily or ineluctably involve that individual in the debate on deployment policies. One need only suppose the situation where a political candidate announced to his colleagues, "I am an adulterer." Surely this candidate would not have any First Amendment right to preclude that statement operating as a de facto disqualification for employment as the President of the United States. Thus, the candidate's statement of a "fact private in nature" might have a very public effect. Still, this effect would not be part of any "unfettered interchange of ideas for the *bringing about of political and social changes desired by the people.*"[88] The statement "I am an adulterer" or "I am a homosexual" does not—either expressly or by implication—*advocate* anything. Therefore, it is illogical to ascribe a place for it in public debate.

Thus, because statements of identity—"I am a homosexual"—are not part of an "unfettered interchange of ideas for the bringing about of political and social changes desired by the people," declarations of one's homosexuality are not protected speech.

The homosexual exclusion policy as an abridgement of protected speech

Even if the statement "I am a homosexual" were protected speech, there must be an actual abridgement of the First Amendment guarantee of free speech before the constitutionality of the regulation is jeopardized. To demonstrate this abridgement, the court must first precisely determine what the regulation regulates. If speech is implicated, the individual's interest in making the speech must, in essence, be balanced against the governmental interests furthered by the regulation. Clearly, "a sufficiently important governmental interest . . . can justify incidental limitations on First Amendment freedoms."[89] In Rich v. Secretary of the Army, the court held "any incidental effect that the Army policy of excluding homosexuals has on First Amendment rights is justified by the special needs of the military. To insure that the armed services are always capable of performing their mission, the military 'must insist upon a respect for duty and a discipline without counterpart in civilian life.'"[90]

The question of precisely what the homosexual exclusion policy regulates is probably the most confounding of any raised throughout this litigation. The question requires choosing between two opposite explanations of how the policy works. These opposite explanations are:

The homosexual exclusion policy regulates *speech* (the declaration, "I am a homosexual") and therefore *homosexual status* (the psychosociological content, or import, of that speech) that is reasonably inferred from that speech.

The homosexual exclusion policy regulates *conduct* (homosexual manifestations) and that conduct—in the past, present, or future—is reasonably inferred from the declaration, "I am a homosexual."

Resolving the conflict between these two explanations depends, in part, on whether it is theoretically or practically possible to create a strict dichotomy between *status* and *conduct*.

No plaintiff challenging the homosexual exclusion policy claimed the Uniform Code of Military Justice proscription of sodomy was unconstitutional.[91] Few plaintiffs disputed that the regulatory policy could be applied against them and that discharge was permissible if they were shown to have committed homosexual acts.[92] The controversy, then, centered on the inference to be drawn from one's admission of homosexuality, without other evidence of homosexual acts.

Plaintiffs argued an admission of homosexuality is wholly unrelated to conduct. Such an admission, according to plaintiffs, signifies one's homosexual status, but fails completely to indicate or even suggest the possibility of past, present, or future homosexual conduct. Pursuing this argument—that status does not suggest conduct—to its logical conclusion often required dogged persistence and drastic maneuvers. In *Steffan*, for example, plaintiff's counsel informed the court:

> Your honor, . . . [t]hey're trying to turn a *status* case into a *conduct* case. . . . If [plaintiff] is asked at a deposition have you ever engaged in [homosexual] conduct I'm going to direct him not to answer. . . . [Plaintiff] refused to answer any questions related to whether he engaged in homosexual acts while attending the [Naval] Academy . . . on the grounds of relevance and fifth amendment privilege.[93]

The outcome of the above discussion was that the district court dismissed for failure to cooperate with discovery. The Navy argued plaintiff's declaration of homosexuality put homosexual conduct in issue. On review, the court of appeals held that plaintiff had been excluded from the military "solely because of his 'status' as a homosexual." Therefore, the court found, plaintiff's declaration of homosexuality was irrelevant to homosexual acts as a matter of law.[94] Thus, the court of appeals fashioned a strict dichotomy between status and conduct. Steffan had resigned from the Navy before investigators completed their inquiry into his participation in homosexual acts.[95]

While plaintiffs took care to avoid "turning a status case into a conduct case" and maintained that a declaration of homosexuality had nothing to do with homosexual acts, the Army, on the other hand, argued that common sense simply told otherwise.[96] Common sense may seem a pedestrian basis for such a critical argument, but it is an important and recognized commodity in the law and should not be underrated as a factor in decision-making.[97] At oral argument in *Ben-Shalom*, for example, the Court of Appeals for the Seventh Circuit put several common-sense questions to counsel.

Does the Army have any way of distinguishing a practicing homosexual from an admitted homosexual?

Can you be a homosexual and have no propensity to commit homosexual acts?

How does one know he or she is a homosexual?

Isn't experience—acts—the most telling indication that you are a homosexual?[98]

These questions highlight the analytical difficulty in bifurcating status and conduct—a task the Army declined in any event. In the Army's view, an obvious connection exists between an admission of homosexuality and the propensity to commit homosexual acts over time. Thus, the homosexual exclusion policy treats an admission of homosexuality as raising a rebuttable presumption of homosexual conduct, either in the past, present, or future.

Indeed, the policy explicitly provides the soldier with the opportunity to rebut the presumption of homosexuality drawn from homosexual acts or declarations of homosexuality.[99] The court in *Ben-Shalom* noted the notice of intent to discharge Ben-Shalom from the Army advised:

her admission of homosexuality gave rise to a presumption that she was a homosexual, and that therefore she had thirty days within which to submit a response rebutting that presumption. In her response to the Army's notice, plaintiff again admitted that she is a lesbian.[100]

In challenges to the homosexual exclusion policy, courts often turned their decisions on whether they accepted or rejected the neat dichotomy between homosexual status and homosexual conduct urged by plaintiffs.

Homosexual status as the focus of the homosexual exclusion policy. Plaintiffs challenging the homosexual exclusion policy are not the first to allege a status-conduct dichotomy. The metaphysical distinction, however, between what a person *is* and what a person *does* remains difficult to articulate, much less to account for in the broad sweep of regulations such as military personnel policies. The Supreme Court, in Robinson v. California,[101] confronted this dichotomy when California enforced a statute that made being a narcotics addict a crime. A close analysis of *Robinson* is instructive.

The Supreme Court first found the statute at issue in *Robinson* could have been construed to reach conduct, rather than status. The Court, however, was bound by the fact California explicitly had construed the statute as criminalizing the status of being addicted to narcotics.[102] The evidence presented, and found sufficient, at Robinson's criminal trial for the offense of addiction was physical evidence of drug use ("tracks") and the defendant's statement that he used drugs. The Court did not discount that these facts constituted proper evidence of drug *use* (conduct). It rejected, however, the contention the evidence circumstantially proved *addiction* (status).[103]

The Court held the statute was constitutionally infirm because it punished an individual for a status—that is, being addicted to drugs—that could be acquired without engaging in any criminal behavior.[104] Likewise, plaintiffs argue, the homosexual exclusion policy is infirm because it punishes an individual for a status—that is, homosexuality—that can be acquired without engaging in any criminal behavior. This argument—whatever its merits—does not follow logically from *Robinson*.

The homosexual exclusion policy is distinguishable from the statute in *Robinson* in several important respects. First, operation of the homosexual exclusion policy is not similar to a criminal prosecution. Conviction of a criminal offense requires proof *beyond a reasonable doubt*. That standard is the highest standard of proof known to the law. It is vastly, and rightly, different from the Government's burden in administrative proceedings or policy determinations where myriad factors—such as cost and administrative burden—are proper considerations.[105]

Indeed, "not only does the standard of proof reflect the importance of a particular adjudication, but it also serves as '*a societal judgment about how the risk of error should be distributed between the litigants.*'"[106] In other words, the function of the standard of proof is to instruct the fact finder on the degree of factual correctness society demands in light of the consequences of the determination.[107] The consequence of the homosexual exclusion policy is simply denial of the opportunity to be a soldier.

This highlights the second way in which the homosexual exclusion policy is different from the statute in *Robinson*. Though denial of the opportunity to be a soldier may cause personal disappointment, exclusion from military service is not punishment any more than exclusion from an entitlements scheme is punishment.[108] The *Rich* court explained that when a soldier is discharged under the homosexual exclusion policy, "[n]o punishment is involved . . . the Secretary has the statutory authority to decide that [homosexuality] make[s] one ineligible for military service."[109]

Even if the homosexual exclusion policy implicates status and not conduct, the Court in *Robinson* explicitly recognized administrative and penal sanctions based solely on status sometimes are permissible and appropriate. The Court discussed the parameters of the State's police power and responsibility regarding public health and welfare and observed:

[A] State might establish a program of *compulsory treatment* for those addicted to narcotics. Such a program of treatment might require periods of involuntary confinement. And *penal sanctions* might be imposed for failure to comply with established compulsory treatment procedures. . . . A State might determine that the general health and welfare require . . . compulsory treatment, involving quarantine, confinement, or sequestration.[110]

Finally, the distinction between the statute in *Robinson* and the homosexual exclusion policy is seen most clearly by analogy. If a person presented himself for enlistment in the Army and declared, "I am a drug addict"—the *status* at issue in *Robinson*—the Army certainly and constitutionally could exclude that individual from military service. This is true even without independent evidence of the individual's past, present, or possible future drug use, and even though "a person may even be a narcotics addict from the moment of his birth"[111] and addiction may be at "the core of one's personality."[112]

Thus, even if the homosexual exclusion policy did somehow implicate status and not conduct, that would not necessarily render the policy unconstitutional.[113]

Homosexual conduct as the focus of the homosexual exclusion policy. For the most part, the military refused the invitation to step into the abstract status-conduct quagmire. Instead, the Army asserted the common sense proposition that when a person says, "I am a homosexual," his or her declaration—as is usual throughout the law—should be given its ordinary meaning. The court in *Ben-Shalom* took this common sense approach and observed, "it cannot be said to be without individual exceptions, but [a declaration of homosexuality] is compelling evidence that plaintiff has in the past and is likely to again engage in [homosexual] conduct."[114]

The Army regulation defines a homosexual as a person who "engages in, desires to engage in, or intends to engage in homosexual acts."[115] Homosexuals have defined themselves as individuals "who ha[ve] committed sodomy."[116] In fact, one plaintiff, in a twist rejecting the status-conduct dichotomy, argued "homosexual status is accorded to people who engage in homosexual conduct."[117]

In another twist, one judge who based his dissent on a status argument, nevertheless asserted he would "be the first to admit that homosexuals, in sexually expressing their affection for persons of their own sex, frequently engage in sodomy."[118] Another court held that to call someone a "queer" imputes the act of sodomy to him.[119] One homosexual activist group, in fact, calls itself "Queer Nation." Researchers have noted "homosexual men tend to define themselves in sexual terms (the very name, homosexual, defines them as having sex with persons of the same gender)."[120] Finally, the American Heritage Dictionary defines homosexuality as "sexual desire for others of one's own sex [and] sexual activity with another of the same sex."[121] Plainly, the ordinary meaning of a declaration that one is homosexual must embrace at least some aspect of sexual activity, desire, or intent.

Plaintiffs challenging the homosexual exclusion policy based on First Amendment guarantees of freedom of speech generally had made forthright, unequivocal declarations they were homosexual.[122] In *Ben-Shalom*, the court found such a declaration, "if not an admission of [homosexual] practice, at least can rationally and reasonably be viewed as reliable evidence of a desire and propensity to engage in homosexual conduct."[123]

Thus, the declaration has legal significance because *it is evidence relevant to conduct*. This is true of *any* statement of identity. In Harper v. Wallingford, for example, an inmate was denied access to a North American Man-Boy Love Association (NAMBLA) newsletter. Prison officials testified, and the court accepted, that "although it was *not clear* the plaintiff is a pedophile, the fact that he is an anti-social personality and he has sodomized a child in the past make it more likely that he will commit such an act in the future."[124] The court held it was permissible to deny plaintiff access to the newsletter.

Now suppose plaintiff declared "I am a pedophile"—in other words, he made it *perfectly clear* he was a pedophile. In this instance, the prison's evidence for withholding the NAMBLA materials would be even stronger. The officials reasonably could conclude the statement "I am a pedophile", even "if not an admission of [pedophilic] practice, at least can rationally and reasonably be viewed as reliable evidence of a desire and propensity to engage in [pedophilic] conduct."[125]

Courts, therefore, rejected arguments that the homosexual exclusion policy implicated only a "psycho-sociological status." The homosexual exclusion policy is, as the law clearly recognizes, ultimately directed at controlling homosexual conduct—criminal and otherwise—within the military. And "controlling" conduct, quite logically, is not limited to taking action after the conduct has occurred. Controlling conduct includes taking measures to avoid the adverse impact of *potential* conduct. As one court noted,

it makes little difference whether a person has committed homosexual acts, or would like to do so, or intends so to do. A person in one of the last two categories could reasonably be deemed to be just as incompatible with military service as one who engages in homosexual acts. Certainly, the morale factor could reasonably be considered to be the same, and the Army understandably would be apprehensive of the prospect that desire or intent would ripen into attempt or actual performance.[126]

In sum, "absolutely no First Amendment violation" occurs when a declaration of homosexuality is used as a basis for discharge from the military, even without other evidence of homosexual acts.[127] The *Ben-Shalom* court stated the relative positions of the parties under the regulation as follows: "Plaintiff can say what she wants to say about homosexuals, but if plaintiff admits she is one, then the Army has the right to say something in response."[128]

The military has concluded that a statement of homosexuality is reliable evidence of a propensity for homosexual conduct over time. Thus, the military has declined to "assume the risk that the presence of homosexuals within the service will not compromise the admittedly significant government interests asserted in the [homosexual exclusion policy]."[129]

The scope of and authority for the homosexual exclusion policy

The last prong of the speech analysis appears in United States v. O'Brien.[130] *O'Brien* sets out a test that evaluates the authority for and importance of the government's underlying interest in a regulation, as well as the scope of the regulatory restriction—if any—on First Amendment rights.[131] As demonstrated above, the homosexual exclusion policy does not implicate First Amendment rights. Assuming it did, however, the homosexual exclusion policy comports with *O'Brien*.

Subsequent sections will discuss the governmental interests underlying the homosexual exclusion policy in detail. Courts readily have accepted the proposition that the policy decision on homosexuality is within the inherent authority of the executive and legislative branches to raise armies.[132] Courts also found without difficulty that the policy is supported by substantial and important governmental interests related to the combat readiness of the armed forces.[133] Based on the types of regulatory restrictions on First Amendment rights of soldiers already upheld by the Supreme Court, finding the homosexual exclusion policy swept with too broad a brush would have been surprising indeed.

In Brown v. Glines, for example, the Court upheld a regulation prohibiting a soldier from circulating petitions addressed to the Congress, even though a soldier has a statutory right to communicate directly with members of Congress.[134]

In Goldman v. Weinberger, the Court upheld a regulation that precluded an orthodox Jew from wearing a yarmulke while in military uniform even though the right to the free exercise of religion is explicitly established and protected by the First Amendment.[135]

Finally, in *O'Brien*, the Court upheld a statute that imposed criminal sanctions for destruction of draft registration cards even though an established right to engage in symbolic, political speech exists.[136] Even if declarations of homosexuality were protected speech, exclusion from military service does not approach the severe result of criminal sanctions found valid in *O'Brien*.

The Supreme Court repeatedly has recognized that, in the military context, First Amendment freedoms sometimes must be compromised.[137] Military courts also have addressed this point frequently. In United States v. Priest, the Court of Military Appeals discussed the operation of the First Amendment within the military and noted,

In military life . . . other considerations must be weighed. The armed forces depend on a command structure that at times must commit men to combat, not only hazarding their lives but ultimately involving the security of the Nation itself. Speech that is protected in the civil population may nonetheless undermine the effectiveness of response to command. If it does, it is constitutionally unprotected.[138]

Plainly, "no [soldier] . . . has a right, be it constitutional, statutory, or otherwise, to publish any information which will imperil his unit or its cause."[139] In the military, First Amendment rights may be limited commensurate with the need for discipline, and "there is no constitutional right to be free from an appropriate degree of discipline, if one is affiliated with an organization where discipline is necessary."[140] Thus, even if the homosexual exclusion policy implicated First Amendment rights, the government's interest in good order, discipline, and morale is paramount.

In challenges to the homosexual exclusion policy, courts declined to create "a First Amendment 'exclusionary rule' to bar the use of [one's] statements as evidence of [one's] homosexuality."[141] This reference to the evidentiary exclusionary rule based on the Fourth Amendment is enlightening. At a criminal trial, where the standard of proof is vastly higher than in administrative proceedings, a statement of one's homosexuality could be admitted as evidence relevant to an element of proof—for example, as a fact relevant to nonconsent in a rape case[142] or to corroborate a confession to homosexual acts.[143] It simply does not make sense, therefore, to argue the law prohibits use of the same statement of homosexuality as relevant evidence in an administrative proceeding.

Thus, whether the statement "I am a homosexual" is speech— even protected speech—it is, like all other statements of identity, a permissible basis for application of military personnel policies, including military exclusion.

HOMOSEXUALITY AND FIRST AMENDMENT RIGHTS:
Conclusion

The homosexual exclusion policy operates in the same way as other military exclusion policies. It is not unique among policies setting standards for the recruitment and retention of soldiers. All military personnel policies have the same goal—that is, to compose a fighting force that has the best chance of achieving the highest standard of combat readiness. All military personnel policies are premised on the certain fact that "the essence of military service is the subordination of the desires and interests of the individual to the *needs of the service*."[144]

The homosexual exclusion policy does not implicate, much less violate, First Amendment guarantees of rights of free association or free speech. If it did, most other military exclusion policies likewise would be infirm. Moreover, if a statement of one's homosexuality were protected by the First Amendment, this would give special significance to the homosexual identity as opposed to other identities that are service-disqualifying. In the absence of a principled way to distinguish homosexuality from other service disqualifiers—much less from other sexual preferences—that result is unwarranted in law and logic.[145]

CHAPTER THREE

THE HOPE AND THE GLORY

The Art of Styling Constitutional Claims

OVERVIEW OF THE LITIGATION, 1971–1991

In challenges to the homosexual exclusion policy based on theories of privacy and First Amendment rights, social commentators and even some legal analysts have too often equated controversy with substance—taking the position that "where there is smoke, there is fire," or "where there are controversial spoken words or sexual acts, there is constitutional infirmity" in any and all government actions based on sexual acts.

However, challenges to the homosexual exclusion policy based on the "due process" and "equal protection" clauses have not led the courts to such a conclusion. Instead, these cases have tended to focus on a distinction between "status" and "conduct," between what a person is *inclined* to do as opposed to what a person actually *does*.

In litigation over this question, plaintiffs have tended to focus their arguments on the meaning of language rather than on more substantive matters. Consequently, this focus requires a particularly close analysis of the way in which opponents of the policy have styled or stated their claims and how these claims are likely to be styled in the future. This chapter will provide the necessary close

31

analysis and review of the law pertaining to "due process" and "equal protection" guarantees.

HOMOSEXUALITY AND "DUE PROCESS"
Substantive and Procedural Guarantees

In 1986, the Supreme Court held in Bowers v. Hardwick that there was no fundamental right to engage in sodomy.[1] *Hardwick* was decided on a "substantive due process" analysis.[2] As a legal concept, substantive due process is as hard to define as the concept of "penumbral rights" surrounding explicit rights in the Constitution. Indeed, no discussion of substantive due process can proceed without referencing the "historically recurrent debate over whether 'due process' includes substantive restrictions"[3]—in other words, the polite and lawyerly debate over whether "substantive due process" as a legal principle even exists.

The rationale for "substantive due process" has been stated as follows:

[T]he full scope of the liberty guaranteed by the Due Process Clause cannot be found in or limited by the precise terms of the specific guarantees elsewhere provided in the constitution. This 'liberty' is not a series of isolated points . . . It is a rational continuum which, broadly speaking, includes a freedom from all substantial arbitrary impositions and purposeless restraints.[4]

Because substantive due process recognizes rights that are otherwise unenumerated, a claim under substantive due process must involve a right that could be classified as fundamental.[5] And how do we know what the fundamental nature of a right is? We find it in historical inquiry. In *Hardwick*, for example, Chief Justice Burger inquired of history whether sodomy was a fundamental right. The Chief Justice found:

Proscriptions against [sodomy] have ancient roots. . . . [T]o claim that a right to engage in such conduct is "deeply rooted in this Nation's history and tradition" or "implicit in the concept of ordered liberty" is, at best facetious. . . . To hold that the act of homosexual sodomy is somehow protected as a fundamental right would be to cast aside millennia of moral teaching.[6]

Indeed, it is fair to say historical inquiry is the *only* guideline in analyzing substantive due process claims. As Justice Scalia deftly has observed,

> [When the applicable principles are not] set forth in the Constitution [they are not] known to the nine Justices of this Court any better than they are known to nine people picked at random from the Kansas City telephone directory . . . [but it] is at least true that no "substantive due process" claim can be maintained unless the claimant demonstrates that the State has deprived him of a right *historically* and *traditionally* protected against State interference.[7]

In a substantive due process challenge to the military's single parent exclusion policy, the court upheld the policy, finding it was "facially neutral and had a strong historic and rational basis."[8]

From time to time courts have attempted to refine the analysis for substantive due process claims, most notably in cases challenging the homosexual exclusion policy. In Beller v. Middendorf, the court offered the opinion that "recent decisions indicate that substantive due process scrutiny of a government regulation involves a case-by-case balancing."[9] The court called this standard of review *"heightened solicitude,"* and then found the homosexual exclusion policy did not violate substantive due process.[10] Another court reviewed a due process challenge to the federal government's homosexual exclusion policy on a standard of *"inherent absurd[ity]"*.[11]

In addition to the "heightened solicitude" and "inherently absurd" standards, one judge applied an *impact-on-the-individual* standard of review to substantive due process claims.[12] This review focused "on the significance and intimacy of a personal decision *to the individual."*[13]

This focus, like "heightened scrutiny" and "inherent absurdity," is quite different from historical inquiry. Moreover, the impact of a

regulation on an individual is not even relevant to military personnel policies. Military personnel policies focus, as they must, on the
needs of the service.[14]

The Supreme Court has always returned to the question of
whether or not the "right" under scrutiny is "implicit in the concept
of ordered liberty."[15] As already noted, this inquiry is essentially an
historical one and most courts realize this truth. Thus the claim of
rights in homosexuality is informed by history: "[i]n view of the
strong objection to homosexual conduct, which has prevailed in
Western culture for the past seven centuries, we cannot say that [the
proscription of sodomy] is 'totally unrelated to the pursuit of' . . .
implementing morality, a permissible state goal."[16] One commentator notes that sanctions for homosexual acts have been recorded as
early as the year 1444.[17] The 1916 Articles of War proscribed sodomy and the military has had some form of official homosexual
exclusion policy since at least the 1940's.[18]

In challenges to the homosexual exclusion policy, most courts
recognized the whole tenor of Supreme Court case law counseled
the greatest caution in expanding or creating fundamental rights.[19]
Nevertheless, at least one judge based his legal analysis on his view
that,

> Society's attitudes toward various aspects of sexuality and personal
> autonomy have changed enormously in recent years. As *social issues*
> have become *legal issues*, the Supreme Court has redefined the bounds
> of the government's power to prohibit certain activities or types of
> behavior.[20]

There is little to support the view that the Supreme Court lately has
embarked on any notable redefinitions of government power. The
statement that society's attitudes have changed enormously from
those historically embodied in the law is also problematic.[21]

Even before *Hardwick*, courts addressing substantive due process
challenges to the homosexual exclusion policy refused to find a
constitutionally protected right to engage in homosexual acts.[22]
Since no fundamental right triggered heightened scrutiny, the pol-

icy was tested against the rational basis standard.[23] Courts routinely held that the policy meets this test without difficulty.[24]

Besides substantive due process claims, plaintiffs urged procedural due process claims. To raise an issue of procedural due process, a claim must implicate a property or liberty interest. For example, one plaintiff, denied foreign service because of his homosexuality, did not have a procedural due process claim because no liberty interest was implicated by the mere loss of some employment opportunities.[25]

Process is constitutionally due only after an identifiable interest is implicated. In Doe v. Casey, plaintiff challenged his exclusion from the CIA based on his homosexuality. The court held "even if the CIA deprived Doe of his liberty interest in his reputation . . . our inquiry is not at an end. The due process clause requires that the CIA not deprive Doe of his liberty interest without due process of law."[26]

It is well settled a soldier has no property interest in continued military service and enlistment itself is a privilege, not a right.[27] In Ben-Shalom I, the district court held plaintiff "cannot establish any 'entitlement' to continued service sufficient to elevate her subjective expectancy to a status worthy of due process protection. . . . When she enlisted [she] was 'entitled' to be retained only for as long as she complied with Army regulations."[28]

If the law were otherwise, every person who had ever been denied enlistment—and every soldier who had ever been denied reenlistment—would have a potential due process claim. In Lindenau v. Alexander, the court noted if it ruled single parents had a property interest in enlistment, then this interest "could result in a flood of litigation attacking all enlistment regulations and criteria."[29] Likewise in Cortright v. Resor, the Court refused to fashion a rule giving every soldier a "ticket to the courthouse to challenge any [decision] that was distasteful to him."[30]

Homosexual plaintiffs, like single parents or other excluded plaintiffs, may have an "abstract need or desire for" military service.[31] This need or desire, however, is not sufficient to provide the "legitimate claim of entitlement" required for constitutional protection to obtain.[32] Plaintiffs' claims to a legitimate entitlement to military service were complicated by the fact many plaintiffs entered the military by evading the homosexual exclusion policy.[33] Thus, they were not eligible to enter military service in the first instance, much less entitled to remain in service in violation of personnel regulations. As the Supreme Court observed in Bowen v. Gilliard, "some perspective on the issue is helpful here."[34] Applying the *Gilliard* analysis, it is clear that:

> Had no [plaintiff ever served in the military], and had [the military] then instituted a [homosexual exclusion policy], it is hard to believe that we would seriously entertain an argument that the new [policy] constituted a taking. Yet, somehow, once [homosexuals are in the military] and [the military] sees a need to [discharge] them . . . , the "takings" label seems to have a bit more plausibility. *For legal purposes though, the two situations are identical.*[35]

Thus, no property interests are implicated by the homosexual exclusion policy.[36]

"Liberty interests," likewise, are not proved by subjective need or desire. The requirements for finding a protected "liberty interest" in government employment are the discharge or termination of a government employee based on false and stigmatizing reasons and the subsequent publication of those reasons by the employer.[37]

In an attempt to demonstrate that the homosexual exclusion policy implicated a protected liberty interest, plaintiffs advanced two theories of stigma. The first theory asserts that the very fact of homosexuality—or being labeled a homosexual—is stigmatizing. One lower court accepted this theory and found, "rightly or wrongly, a disclosure of homosexual activity will tend to stigmatize a person, particularly when coupled with an involuntary separation from military service."[38] Nevertheless, since most plaintiffs were self-identified and admitted homosexuals, courts readily rejected this theory.[39]

The second theory asserts that discharge from the military is itself stigmatizing, because it brands a person as "unfit" to soldier.[40] In Neal v. Secretary of the Navy, for example, plaintiff's service record indicated homosexual acts, but he was discharged for the convenience of the government. Plaintiff argued he was "branded as being unfit to be a Marine," and thus had a liberty interest implicated in his military discharge.[41]

This theory also was rejected.[42] The *Beller* court found the basis for discharge under the homosexual exclusion policy was homosexuality, not fitness for service. Thus, there could be no stigma for unfitness "since the regulations do not make fitness of the particular individual a factor in the decision to discharge."[43]

In sum, the homosexual exclusion policy implicates no protected property or liberty interest. Thus, the policy does not raise procedural due process claims on those grounds.[44] Further, the homosexual exclusion policy does not infringe any fundamental right. Thus, substantive due process likewise provides no basis for invalidating the policy.

HOMOSEXUALITY AND "EQUAL PROTECTION"

"Equal protection" challenges to the homosexual exclusion policy have engendered lively debate and still are offered as the most viable route to judicial invalidation of the homosexual exclusion policy.[45] Equal protection analysis is well settled. It entails three tiers: the "rational basis" test for regulations that do not violate a fundamental right or burden a "suspect" or "quasi-suspect" class; heightened scrutiny when a "quasi-suspect" class is burdened; and strict scrutiny when a "suspect class" is burdened.[46] The equal protection analysis determines the appropriate standard of review to apply, not the constitutionality of the policy in question. This section will discuss applicability of the various standards of review in terms of homosexuals as a "suspect class" and homosexuality as a "fundamental right."

WHAT IS A "SUSPECT CLASS"?

In High Tech Gays v. Defense Industrial Security Clearance Office, the court reiterated the prerequisites to heightened scrutiny under equal protection guarantees. The court stated,

[t]o be a 'suspect' or 'quasi-suspect' class, homosexuals must 1) have suffered a history of discrimination; 2) exhibit obvious, immutable, or distinguishing characteristics that define them as a discrete group, and 3) show that they are a minority or politically powerless, or alternatively show that the statutory classification at issue burdens a fundamental right.[47]

Although it often has been argued otherwise, mere exclusion from a government program or service is not a factor in establishing a suspect classification. In *Gilliard*, for example, the Supreme Court held that families excluded from Federal Aid to Families with Dependent Children program were not a suspect class.[48]

Most courts have accepted the argument that homosexuals have suffered a history of discrimination.[49] Public controversy, however, exists on this point. In a debate on a legislative proposal regarding homosexual rights, one lawmaker explained,

"I just don't think that civil rights spring from a private social activity." [The lawmaker] . . . also said that he did not believe the gay community has demonstrated that a pattern of discrimination exists against homosexuals. "Every other civil rights movement showed there was a pervasive pattern of discrimination. I don't think that pattern has ever been established by anyone at all in this debate."[50]

In places where ordinances prohibiting certain types of discrimination against homosexuals have been enacted—generally at the municipal level—the ordinances are rarely invoked.[51] A leaflet supporting a referendum to repeal a sexual preference ordinance in Tampa, Florida, states "homosexuals are far from being an 'oppressed minority.'"[52] The leaflet cites a report in the *Wall Street Journal* showing the average household income for homosexuals was $55,430, while the national average was $32,144, the average in Hispanic households was $17,939, and the average income in Black households was $12,166.[53]

The report also showed among homosexuals, 59.6% were college graduates, compared to a national average of 18%; 49% of homosexuals were in professional or managerial positions, compared to a national average of 15.9%; 65.8% of homosexuals were overseas travelers, compared to a national average of 14%; and 26.5% of homosexuals qualified as frequent fliers, compared to a national average of 1.9%.[54] Comparisons were even more disparate when homosexuals were compared in these categories to Hispanic or Black households.[55]

One analyst quoted in the *Wall Street Journal* said of homosexuals, "You're talking about two people with good jobs, lots of money, and no dependents. This is a dream market."[56] The Tampa citizens group put it this way: "While black Americans suffered demonstrable economic deprivation due to legal patterns of discrimination and segregation, the average homosexual actually enjoys the benefits of 'the American dream' *at a far greater rate* than the rest of the population. Far from living in daily fear that they will lose their jobs due to 'discrimination,' these national statistics show that almost half of all practicing homosexuals hold either professional or managerial positions."[57]

Despite this public controversy, few if any defendants challenged the notion that homosexuals had suffered a history of discrimination, and courts accepted this argument without objection. At the same time, courts recognized not all discrimination was automatically impermissible or necessarily inappropriate. Indeed, several courts concluded "there can hardly be more palpable discrimination against a class than making the conduct that defines the class criminal."[58] Still, for purposes of suspect classification, homosexuals were held to meet the criterion of having a "history of discrimination."[59]

The other criteria for suspect classification—immutability and minority status or political powerlessness—occasioned more debate. Most courts easily disposed of the question of political power by finding that homosexuals in fact were politically powerful.[60] Clearly, homosexuals are not so helpless as to "command extraordinary protection from the majoritarian political process."[61] Indeed, implicit in one court's analysis was the finding that homosexuals were not protected by the Civil Rights Act because homosexuals were not among the groups the government determined *"require and warrant special federal assistance* in protecting their civil rights."[62]

Although homosexuals were found politically powerful, they easily qualified as a minority, thus meeting the third criterion. Suspect classification, then, hinged on homosexuals meeting the criteria for constitutional protection as a discrete group.

The criteria for constitutional protection as a discrete group—even for denomination as a suspect class—are often denoted by the shorthand term "immutability." The standard in its entirety, however, requires "obvious, immutable, or distinguishing characteristics that define [the proposed class] as a discrete group."[63]

In challenges to the homosexual exclusion policy based on suspect classification, immutability usually was styled as a simple issue of whether or not homosexuals could *change* either their behavior or their sexual orientation.[64] The ability or possibility of change in homosexuals often was presented as an issue of the etiology, or origin, of homosexuality.[65] The theory was if homosexuals did not choose to be homosexual, then they cannot change their homosexuality; if they cannot change their homosexuality, then it is immutable; if homosexuality is immutable, then homosexuals are a suspect class.

This logical sequence (which itself is arguably fallacious) misperceives the standard for suspect classification. Taken together with the requirement for "*obvious* and *distinguishing* characteristics that define a *discrete* group"—and compared to the characteristics of established suspect classes—plainly the *changeability* of the condition that defines the class is only one factor in determining whether a classification is suspect.

A one-factor analysis simply is not sufficient, as shown in Schweiker v. Wilson, where the Supreme Court rejected the lower court's finding that, even though the "mentally ill as a group do not demonstrate *all* the characteristics . . . denoting inherently suspicious classifications," heightened scrutiny nevertheless applied.[66] The Supreme Court found mentally ill persons did *not* constitute a suspect class. Further, unlike a focus on the *changeability* of a particular characteristic, the standard for suspect classification considers the nature of a group *in existence*, not how it *came* to exist.[67]

Nevertheless—and even though it was only one factor—immutability was greatly at issue in claims based on homosexuality as a suspect classification. Plaintiffs argued homosexuality was an immutable characteristic.[68] Courts saw it differently. Whether or not homosexuality was immutable, courts found that homosexuality was not immutable in the same way as race, national origin, alienage, and gender.[69]

Indeed, courts rejected the notion that homosexuality was immutable at all, and held it was "*behavioral* and hence . . . fundamentally different from traits . . . which define already existing suspect and quasi-suspect classes. . . . The behavior or conduct of such already recognized classes is irrelevant to their identification."[70] In fact, in *Ben-Shalom I*, the district court accepted plaintiff's claim that "homosexuals, as defined merely by the *status* of having a particular sexual orientation and *absent any allegations of sexual misconduct*, constitute a suspect class."[71] Thus, the court expressed the opinion that homosexuals were a suspect class *so long as* they did not—or were not alleged to—engage in homosexual *behavior*.

Dissenters to the conclusion that homosexuality was behavioral and thus not immutable presented lively arguments. Judge Canby, for example, stated "[i]t is not enough to say that the category is 'behavioral.' One can make 'behavioral' classes out of persons who go to church on Saturday, persons who speak Spanish, or persons who walk with crutches."[72]

This statement is true, but it does not prove behavioral classes *are* classes with immutable characteristics or suspect classifications. In fact, it is settled that the military might exclude—without constitutional moment—"persons who go to church on Saturday,[73] persons who speak Spanish,[74] and persons who walk with crutches."[75] Regardless of whether or not these characteristics are immutable, they may not be compatible with military service.

Immutability, as seen above, is not the sum of the definition of a suspect class, although it was often used that way. Few challenges to the homosexual exclusion policy have specifically addressed suspect classification in other terms.[76] In other contexts, however, courts have discussed whether homosexuals have obvious, distinguishing characteristics that define them as a discrete group. In *Norton v. Macy*, for example, the court commented that homosexuals were impossible to identify.[77] Similarly, in *Cyr v. Walls*, involving a proposed class action, the court held it could not "certify any class limited to gay persons because of lack of identifiability."[78] The court noted "[c]lass existence is complicated . . . by the fact that gay individuals are not readily identifiable, unlike individuals with certain racial or color characteristics, individuals with certain types of surnames, or individuals of a certain physical sex."[79]

This lack of identifiability of homosexuals has been noted outside the courtroom as well. One researcher surveying the needs of homosexual minors reported efforts were hindered by their lack of visibility.[80]

Some homosexuals may be visible, as in Smith v. Liberty Mutual Ins. Co., where plaintiff, a homosexual, claimed he was "discriminated against . . . because as a male, he was thought to have those attributes more generally characteristic of females and epitomized in the description of 'effeminate.' "[81] Although some commentators assert that "[t]he incidence of 'feminized males' or 'queens' . . . is estimated at about ten percent of the male homosexual population,"[82] effeminate males may or may not be homosexual, so this also is not a way of ready identification.

The lack of identifiability perhaps is the clearest argument against finding homosexuality is a suspect classification. This finding makes sense since the Supreme Court has accorded women—who are infinitely more identifiable than homosexuals—only quasi-suspect class status.[83] Indeed, one court held if gender is not a suspect classification, neither is a classification based on choice of sexual partners.[84]

Regardless of the theory, however, the settled legal conclusion is that homosexuals do not constitute a suspect class.[85] Thus, homosexuals join the ranks of other groups that must look first and primarily to the elected legislature—rather than the judiciary—to secure their objectives. These ranks include single parents,[86] transsexuals,[87] undocumented aliens,[88] prisoners,[89] the mentally retarded,[90] youth,[91] and the aged.[92] Like these groups, homosexuals are not a suspect class. Like homosexuals, these groups are—or may be—excluded from military service.[93] In no instance, however, has military exclusion of these groups occasioned heightened—much less strict—scrutiny of the relevant policy. Moreover, if the question of exclusion was to be decided on the equities, homosexuals would be among those least favorably situated to challenge exclusion since this class is defined by behavior that may be criminal.

FUNDAMENTAL RIGHTS

Plaintiffs, in addition to arguing suspect classification, advanced equal protection theories based on fundamental rights pertaining to homosexuality. Since the Fifth Amendment and the Equal Protection Clause of the Fourteenth Amendment impose identical standards, one might have predicted that *Hardwick* would settle the equal protection issues as roundly as it settled due process challenges to the homosexual exclusion policy.[94] The *Ben-Shalom* court recognized this and concluded, "[a]lthough the [*Hardwick*] Court analyzed the constitutionality of the [sodomy] statute on a due process rather than an equal protection basis, *Hardwick* nevertheless impacts on the scrutiny aspects under an equal protection analysis."[95]

Because *Hardwick* was decided on due process grounds, however, some judges left open the possibility that homosexuals might yet succeed on a "fundamental rights" theory under equal protection guarantees.[96] A similar debate surrounded the meaning of the Supreme Court's summary affirmance in Doe v. Commonwealth's Attorney.[97] In that case, ten years before *Hardwick*, the Court affirmed the constitutionality of a sodomy statute. Some argued the Court's affirmance was on procedural grounds and, thus, was not relevant to challenges to the homosexual exclusion policy.

Challenges to the homosexual exclusion policy based on fundamental rights failed. Equal protection claims, however—even those based on fundamental rights—still are urged as a basis for future litigation.[98]

This restyled assertion of equal protection claims is a direct result of *Hardwick* holding that there is no fundamental right to engage in homosexual acts. After *Hardwick*, it was obvious:

If homosexual conduct may constitutionally be criminalized, then homosexuals do not constitute a suspect or quasi-suspect class entitled to greater than rational basis scrutiny for equal protection purposes. The Constitution . . . cannot otherwise be rationally applied, lest an unjustified and indefensible inconsistency result.[99]

Therefore, to advance an equal protection claim after *Hardwick,* plaintiffs must argue that some fundamental right *distinct from conduct* is implicated by the homosexual exclusion policy. Some argue this other right is a fundamental right to *be* homosexual.[100]

Those who argue this theory only generalize about the contours of the fundamental right to be homosexual. Since, in some sense, every person is what he is regardless of law, policy, religious tenets, or public or private opinion to the contrary, the "right to be" is an obscure concept indeed. The most renowned articulation of a right to "be" no doubt is Justice Brandeis' observation that the true balance between society's interests and the interests of the individual is found in "the right to be let alone—the most comprehensive of rights and the most valued by civilized men."[101] It cannot be said that the homosexual exclusion policy abridges any right to be let alone. Indeed, arguably it advances a homosexual's interest in being let alone, since military service imposes a variety of standards of conduct, including the Uniform Code of Military Justice.

An attempt at defining this "right to be" must be made, however, in order to test this right against the constitutional standard of equal protection. After *Hardwick,* any definition of the right to *be* homosexual must avoid the prohibited aspect of homosexual conduct or else be disqualified from constitutional protection. Thus, the concept of the right to be homosexual relies on a strict status-conduct dichotomy. The operation of this dichotomy will be explored in detail, but the simple way to describe the dichotomy is to say that it is embodied in the proposition that what people *are* is separate and distinct from what they *do.*

One commentator, for example, in an attempt to demonstrate homosexual status does not implicate prohibited homosexual conduct, states, "a person can have a homosexual orientation without engaging in proscribed homosexual acts, just as a person can have a heterosexual orientation without engaging in proscribed heterosexual acts."[102]

Proscribed heterosexual acts, of course, are offenses such as rape, adultery, and carnal knowledge. In most instances, heterosexual fornication and sex with one's spouse are not crimes. Thus, the analogy quickly breaks down. A heterosexual can be sexually active—even promiscuous—and not commit proscribed heterosexual acts. A homosexual, however, can avoid proscribed homosexual acts only by abstaining from sexual activity altogether.

Since there is no suggestion that abstinence is part of the bargain in the status-conduct dichotomy, homosexual "status" apparently does not indicate celibacy and, by definition, it must not prove conduct. Thus, the result of the status-conduct dichotomy is the notion of a homosexual as a "psycho-sociological being," a personality simply disembodied from any suggestion of past, present, or future homosexual conduct. Here the homosexual is *only* an identity. That identity has its own vitality and is quite apart—even remote—from its owner's physical acts. It is this homosexual, or this aspect of a homosexual, that arguably has a fundamental right to *be*.

To test the viability of this alleged right, the first question that must be answered is how this fundamental right to be homosexual differs—theoretically or practically—from a right to other states of existence. For example, any one excluded from military service could make a similar claim exclusion impinged his right to be whatever characteristic the Army determined was service-disqualifying. An applicant rejected for being overweight could claim he has a right to be a person who enjoys food. An applicant rejected for adhering to Nazism could claim he has a right to be extreme. Indeed, under the lower court's opinion in *Ben-Shalom I*, the right to "be" would be extended any time governmental regulation implicated—by whatever standard this was determined—"the privacy of the integral components of one's personality—the essence of one's identity."[103]

Even if the law could determine a way to test when some characteristic was integral or essential to one's personality, there is no principled, constitutional way to distinguish between personalities or identities for purposes of affording constitutional protection to some and not others. It hardly would make sense to argue, for example, that a person's interest in his intellectual ability, even if integral to his personality, is so fundamental that it merits constitutional protection from a university's admission policy. Under this theory, the law already allows substantial infringements of one's "identity" because it allows, for example, some infringement on one's practice of religion.[104] This infringement of an explicit and established constitutional right—in some instances an infringement on one's relationship with one's god—may be more critical to one's identity than sexual preference, yet it passes constitutional muster. The particular trait or identity at issue may be at "the core of one's personality, self-image, [or even] sexual identity."[105] Even so, that fact simply does not demonstrate that the trait or identity should "enjoy special constitutional protection"[106] or "be 'afforded shelter' as a fundamental right."[107] Indeed, the opposite conclusion plainly is unworkable, if not unthinkable.[108]

A second inquiry addresses the question of how any right to be homosexual is impinged by the homosexual exclusion policy. This policy denies homosexuals the opportunity to be soldiers. It does not preclude them in any way from being homosexuals.

The Supreme Court's perspective in *Gilliard* is helpful here. In discussing a statutory amendment that excluded some families from welfare benefits, the Court observed, "it is imperative to recognize that the amendment at issue *merely incorporates a definitional element* [regarding who is eligible to enter] into an entitlements program."[109] It is likewise imperative to recognize that the homosexual exclusion policy merely defines who is eligible for military service. All military personnel policies operate similarly. They offer or deny the opportunity to soldier. They neither contemplate nor affect any aspect of a military applicant's right to be whatever he is.

This principle was seen clearly in *Goldman*. There the Court found a regulation constitutionally permissible that directly infringed upon the right to free exercise of religion.[110] Goldman, an orthodox Jew, could continue to follow his religious precepts, but—to the extent they conflicted with military regulations—he had to choose between following his religion and being in the Air Force. The fact that Goldman had to choose did not preclude him from being an orthodox Jew.[111]

As the *Goldman* case demonstrates, all rules can create dilemmas for those who come within their purview. Some conclude that the fact a rule creates unpleasant dilemmas calls into question the soundness of the rule. It is these unpleasant personal dilemmas that the status-conduct dichotomy is designed to fix. In one commentator's view, for example, "[p]eople who know they have a homosexual orientation and who want to serve in the military are faced with a dilemma: disclose and be excluded, or lie and hide."[112] There exists a third, perfectly legitimate choice, however, and that is for homosexuals—and single parents[113] and any one else who is not eligible for military service—to choose a profession other than military service.

The fact that an individual must choose between obeying the rules of an organization and not joining the organization does not impact on one's opportunity to think or feel whatever and however one wishes, and it does not call the soundness of the rule into question. In fact, the soldier always is expected to choose obedience to lawful orders and regulations. The Manual for Courts-Martial plainly states, "the dictates of a person's conscience, religion, or personal philosophy cannot justify or excuse the disobedience of an otherwise lawful order."[114]

A final inquiry must test the logic and common sense of the status-conduct dichotomy itself, because the alleged fundamental right to be homosexual depends upon this dichotomy completely for its legal viability. Can a personality or identity so easily and cleanly be separated from the physical body that houses it?

Close analysis shows—in practical terms—the status-conduct dichotomy is simply ephemeral. The dichotomy is difficult in theory and unrecognizable in reality. To constitute a workable policy premise, the status-conduct dichotomy requires either factual celibacy on the part of the homosexual, or deliberate ignorance of the facts on the part of the Army. The first is unrealistic, the second unacceptable.[115] Thus, the status-conduct dichotomy—and any fundamental rights alleged to arise from it—is not an appropriate conceptual basis for law or policy. Because of the nature of the interests at stake, this is particularly true in the unique dynamic of military personnel policies.

HOMOSEXUALITY AND EQUAL PROTECTION:
Conclusion

In sum, plaintiffs were unable to establish a fundamental right to engage in homosexual acts. Thus, equal protection challenges to the homosexual exclusion policy based on fundamental rights failed. While some argue a fundamental right to be homosexual is protected by equal protection guarantees, the right to "be" is amorphous and there are no apparent principles by which to distinguish between states of being for purposes of constitutional protection. Thus, the law is unlikely to create a new category of equal protection claims based on the right to "be." Because the homosexual exclusion policy does not burden a fundamental right or create a suspect classification, it was tested against the rational basis standard. As courts repeatedly held, the policy meets that standard without any difficulty.[116]

SUMMARY:
Legal Challenges to the Homosexual Exclusion Policy, 1971–1991

Twenty years of litigation on the homosexual exclusion policy has resulted in a substantial and virtually unanimous body of law affirming the policy's constitutionality. The homosexual exclusion policy does not implicate—much less abridge—constitutional guarantees of privacy, free association, free speech, due process, or equal protection. Moreover, the law explicitly and repeatedly recognized that the homosexual exclusion policy has a rational basis. Thus, the first question in reviewing policy—is the policy *permissible?*—is answered soundly in the affirmative. The answer to the next question—is the policy *appropriate?*—takes some lessons from the law, but encompasses a broader and more detailed range of practical factors as well. The next chapter will discuss practical perspectives on whether or not the military has any basis upon which to conclude that the homosexual exclusion policy is an appropriate striking of the balance between individual desires and the needs of the service.

CHAPTER FOUR

———

The Homosexual Exclusion Policy and

the Rational Basis Test

INTRODUCTION
The commander's major task is to fight a war, not a lawsuit.[1]

Though it may have provoked more litigation, the homosexual exclusion policy is not significantly different from other military exclusion policies—or, for that matter, other military personnel policies in
general. Such policies are intended to give America the best possible
fighting force. They are also designed to discriminate between
those classes of individuals whose potential for successful soldiering is strong and those whose potential is weak. A military personnel policy that excludes certain classes of individuals from
soldiering—regardless of the nature of the group excluded—is intended simply to ensure that "[military] personnel can rapidly
respond to national defense requirements and fulfill their duties
within the military community."[2]

These policies are not "anti-single parent," "anti-overweight people," "anti-high school dropout," "anti-people with disabilities," or
"anti-homosexual." Such policies are not anti-*anybody*. They are
simply "pro-military."

Nevertheless, because of the attention the homosexual exclusion
policy has garnered in recent years, it is important to remember the

50

context in which the policy operates. Challenges to the homosexual exclusion policy—like most challenges to military personnel policies—focus on the desires and interests of the individual plaintiff. Despite the individual feelings involved, however, this viewpoint is out of focus because the "essence of military service 'is the subordination of the desires and interests of the individual to the needs of the service.' "[3]

This backdrop is necessary to put the homosexual exclusion policy in practical perspective and to begin to see it as only a small part of the cumulative striking of a balance between individual desires and the needs of the Army. The homosexual exclusion policy has survived judicial scrutiny time and time again, yet it remains a matter of controversy. Much of this controversy results from a failure to apprehend or accept the way military personnel policies operate. More broadly, this controversy results from a fundamental misperception of what the Army is—a political, social, and *constitutional* entity that *has no analogue in the civilian sector.*

To put the homosexual exclusion policy in perspective, this chapter will examine the legal standard of review applied to the policy—the "rational basis" test—in other words, whether the policy is reasonable or else capricious and arbitrary. The discussion will then focus on the Army's arguments in defending military personnel policies and on the plaintiff's arguments challenging the appropriateness of the homosexual exclusion policy. Finally, the chapter will examine the regulatory rationale for the homosexual exclusion policy in detail.

THE RATIONAL BASIS TEST

One of the most important—and most hotly contested—issues raised in challenges to the homosexual exclusion policy has been the appropriate standard of judicial review. As already shown, courts have rejected claims that the homosexual exclusion policy burdens either a fundamental right or a suspect class. Thus, the policy is tested against the "rational basis" standard.

The "rational basis" test has been explained in a variety of ways.

In general, the test operates as a presumption that a government regulation is valid for equal protection purposes *so long as the regulatory classification "rationally furthers some legitimate, articulated state purpose."*[4] Stated conversely, a government regulation is not valid if the regulatory classification is based on "factors which are . . . seldom relevant to the achievement of any legitimate state interest."[5] Thus, the standard is widely recognized as "the most relaxed and tolerant form of judicial scrutiny under the Equal Protection Clause."[6] The judicial deference signified by the rational basis test is grounded in the "constitutional presumption that 'improvident [classifications] will eventually be rectified by the democratic processes.' "[7]

In Dallas v. Stanglin, the Supreme Court reaffirmed that this relaxed and tolerant form of judicial scrutiny was a necessary reflection of the fact that "the problems of government are practical ones and may justify, if they do not require, rough accommodations—illogical, it may be, and unscientific."[8] The Court went on to note that a less flexible approach to determining statutory validity would vitiate "the principle that the Fourteenth Amendment gives the federal courts no power to impose upon the States their view of what constitutes wise . . . social policy."[9]

The homosexual exclusion policy readily survives rational basis scrutiny.[10] From a practical point of view, however, the validity of the rationale underlying the homosexual exclusion policy continues to generate debate. Opponents of the policy argue that the proposition that homosexuality is incompatible with military service cannot be *proved* and therefore is not an appropriate basis for excluding homosexuals from the military. This argument, in essence, is a revisiting of the rational basis issues raised in legal challenges to the homosexual exclusion policy. Thus, the following sections will closely examine the respective litigation positions on whether or not the basis for the policy is "rational."

THE ARMY'S POSITION

Practical as well as legal factors influence the Army's approach to litigation challenging military personnel policies, including challenges to the homosexual exclusion policy. As a practical matter, to comprehend the potential scope of such challenges, it is imperative to understand the scope and breadth of the Army's personnel management operation. This understanding is central to a full appreciation of all that is at stake in such litigation.

Over the last two decades, the United States Army has been at an average strength of nearly *one million soldiers* on active duty. Each year, recruiters screen or process thousands of candidates for military service. Thousands of personnel actions—enlistments, re-enlistments, promotions, discharges, reductions—take place world-wide every day. In addition, when the world changes, the Army changes. This is not a cliche, but rather a serious fact of Army life, and one requiring constant thought and attention, because few social institutions suffer change on the magnitude of America's military.

Army leaders quickly must respond to new missions and emergencies with appropriate and effective measures. In a fighting force of active duty soldiers and reservists that could number in the millions, military personnel management is not a small, insignificant, or easy job. It is a job with late-breaking history always looking over its shoulder. As General Carl Vuono, former Chief of Staff of the Army, noted:

> As we marvel at the collapse of the Soviet empire, we also witness the birth of a new era of uncertainty and peril, an era in which the threats we will confront are themselves ill-defined. . . . [W]e must also prepare for the implications of the instability and chaos that historically trail in the wake of collapsing empires. . . . If the wave of the future is the "come as you are" war, then we must be ready to go at all times.[11]

The magnitude and complexity of managing military personnel —not to mention its critical role in preparing the Army to be ready to go at all times—proves the wisdom that, in the absence of a suspect classification or a burden on fundamental rights, the rational basis test is the appropriate standard against which to review the "rough accommodations" of the practical problems of running an army.[12] The magnitude and complexity of running the U.S. Army—even solely from the viewpoint of personnel management—also explains why, in litigating military personnel policies, as a practical as well as legal matter the Army must hold fast to the rational basis standard of proof.

To prevail on this standard of review, the Army must demonstrate that the questioned "policies and procedures . . . are rationally related to permissible ends."[13] Under the rational basis test as expressed, for example, in Cleburne v. Cleburne Living Center, clearly the Army is not required to establish conclusively that homosexuality is incompatible with military service as a matter of *fact*.[14] Indeed, the term "incompatible" can be a *relative*, rather than *conclusive* term. The Secretary need not determine whether homosexuality is *absolutely* incompatible with military service, *somewhat* incompatible, or so on. The determination simply states the problem: homosexuality is incompatible with military service. The exclusion policy states the solution to that problem, arrived at by the Army's senior military and civilian leadership.

To prevail on rational basis review, "[i]t is enough to show that a *potential* danger exists without the restrictions of the challenged . . . regulation."[15] In Espinoza v. Wilson, for example, plaintiff challenged a prison regulation that permitted withholding of homosexually oriented mail. The court noted,

[P]rison officials had testified that there was a strong correlation between prison violence and inmate homosexuality. . . . The officials also testified that homosexuality played a major role in inmate assaults, rapes, intimidation, extortion and personal abuse. Consequently, the prison officials believe that they could not sanction an activity that *might* further expose innocent inmates to increased dangers, increase the opportunities for identification of homosexual inmates, and increase the ability of dangerous inmates to target homosexual inmates.[16]

The court in *Espinoza* upheld the regulation—thus, permitted withholding literature from the inmates—based on the *potential* danger to discipline within the prison.

The court in *Dronenburg* made a similar finding as to the homosexual exclusion policy when it plainly stated the military "is not required to produce social science data or the results of controlled experiments to prove what common sense and common experience demonstrate."[17] Common sense and experience are—as they should be—integral factors in decisions affecting the composition of the armed forces. The *Dronenburg* court took a common sense approach when it noted, "[t]his very case illustrates the dangers of the sort the Navy is entitled to consider: a 27-year-old petty officer had repeated sexual relations with a 19-year-old seaman recruit."[18]

When the Army is hailed into court to defend a personnel policy, it defends, not the merits of the policy, but rather the Secretary's prerogative to establish the policy and to exercise his professional judgment in the first instance. In essence, the Army defends the Secretary's common sense—a proposition not proven or disproved by social science studies or by what is familiar to trial courts as a "battle of the experts."

These battles have occurred in various lawsuits involving issues relating to homosexuality. In *Aumiller*, for example, "both parties presented expert witness testimony relating to the etiology of homosexuality and its psychological implications."[19] In Baker v. Wade, a psychiatrist and a sociologist testified as experts on homosexuality.[20] Further, a theologian and professor of the New Testament testified, "in his expert opinion, the Bible does not condemn consensual homosexual conduct."[21]

Needless to say, a battle of experts cannot win the day in court because, in any given lawsuit, experts can be found to support the position of each party—not because opinions can be bought, but precisely because every issue has complexities, nuances, and interpretations about which people can disagree. In the end, the decision-maker—whether judge or Secretary—must make a "judgment call" to resolve the issue.

In setting enlistment or retention standards, the Army need not come up with better, more substantiated answers than medical or social science has produced. It is enough that a characteristic, circumstance, or propensity for particular conduct raises a doubt about the potential for successful soldiering. The Army, for example, could look at a soldier's history of adverse reactions to vaccination and then determine that the risk of a future adverse reaction made the soldier unfit for military service—even though some medical experts expressed the opinion that the soldier might be vaccinated safely.[22] Likewise, the Army could determine pedophilia raised a doubt about the potential for successful soldiering—regardless of the fact that experts testified variously on the issue of the relationship between mental status and pedophilia.[23] And a decision-maker would be entitled—even duty-bound—to resolve those doubts in favor of the Army.

In short, a full array of experts—in psychology, sociology, sexology, theology, or any other number of fields—could provide ample expertise to the court. Yet these experts would remain unable to answer what is quintessentially the military's judgment call on which individuals or groups—*considering the administrative burden the Army is willing to bear*—have strong potential for successful soldiering and which do not.[24]

Although over the years an inordinate amount of publicity has attended challenges to the homosexual exclusion policy, it has been imperative for the Army to defend those cases in the same way it has defended lawsuits on other personnel policies. Certainly, the emotional and political dynamics surrounding challenges to the homosexual exclusion policy were quite different than those surrounding, for example, litigation on the single parent exclusion policy. But the legal issues, the standard of review, and the standard of proof were the same in both instances.

In some of the challenges to the homosexual exclusion policy, the Army presented evidence of the regulatory rationale as stated in the regulation itself.[25] In other cases, affidavits or testimony was presented.[26] Some judges rejected both approaches, finding the Army's evidence was "merely a series of platitudes."[27] In the end, however, courts had no difficulty finding that the military interests underlying the homosexual exclusion policy—good order, discipline, and morale—were substantial.[28] The court in *Ben-Shalom* readily recognized the consequential context of the policy and noted:

> [T]he military establishment is very different from civilian life. When necessary, the military must be able to protect and defend the United States. That is a most important government mission, a difficult, demanding and complex one. It requires a trained professional force of reliable, loyal, and responsive soldiers of high morale, with respect for duty and discipline, soldiers who can work together as a team to accomplish whatever missions they may be given by their commanders.[29]

Further, courts found specifically the military's concerns about homosexuality were "not conjectural, but had a basis in fact."[30]

Once the Army presents evidence that there is a rational basis between excluding homosexuals and the maintenance of morale, good order, and discipline within the armed forces, the burden shifts to plaintiffs to show that the proffered basis—and hence the secretarial determination homosexuality is incompatible with military service—is irrational, or is at least an insufficient showing of rationality.[31]

Few plaintiffs have presented affirmative evidence in an effort to negate the Army's showing. In general in these cases, the Army has presented evidence of the effect of *homosexuality* within the military.[32] Plaintiffs have presented evidence of their *individual records* of military service.[33] In some instances, witnesses have testified that plaintiff's homosexuality has not presented problems, but that homosexuality in general could have an adverse impact within the armed forces.[34]

No plaintiff has yet attempted to show that homosexuality *is* compatible with military service.[35] Nevertheless, the proposition that the homosexual exclusion policy is irrational has often been advanced by vigorous argument.[36]

PLAINTIFFS' ARGUMENTS ON THE IRRATIONALITY OF THE HOMOSEXUAL EXCLUSION POLICY

INTRODUCTION

Before turning to the various arguments plaintiffs made in an attempt to show the homosexual exclusion policy is irrational, it is important to set out the fundamentally different perspectives of the parties to such a lawsuit. Lawsuits center on individuals. Lawsuits—in the most basic sense, and however ultimately styled in terms of legal claims—begin with an individual who *feels wronged*. From this perspective, the focus is on the "significance and intimacy of a personal decision [about homosexuality] to the individual."[37] Thus, lawsuits—and the larger social controversy they may reflect—have their origin in individual equities.

In stark contrast to a focus on individual equities, military personnel policies have their origin in the military's weighty obligation to maintain an "efficient and easily administered system of raising armies"[38] whose "primary business . . . [is] to fight or be ready to fight wars should the occasion arise."[39] It is beyond cavil that,

> [m]ilitary regulations must be considered in the light of military exigencies, "must be geared to meet the imperative needs of mobilization and national vigilance . . . and great and wide discretion exists in the . . . interpretation in such matters as what constitutes 'for the good of the service.' "[40]

The "good of the service," naturally, is a much broader concept than the good of any one individual. To analyze what is in the best interest of the service, factors must be considered that have nothing to do with the individual—indeed, factors that may even seem mundane, remote, bureaucratic, or petty to him, as is the case in many government policy choices.

In Mathews v. Eldridge, for example, the Court found that one factor in due process analysis was "the Government's interest, including the *function involved* and the *fiscal and administrative burdens*" entailed in a policy choice.[41] Fiscal and administrative burdens are heavy burdens on an army. And homosexuals—unlike the plaintiff in *Mathews*, who had an obvious property interest in a terminated benefit (Social Security payments)—do not have a property interest in military service. Thus, arguably—and on balance—even greater weight should be accorded to "fiscal and administrative burdens" than would be imposed on the Army by suggested changes to the homosexual exclusion policy. Simply put, the overriding principle in military personnel policies is "the good of the service." This principle clearly cannot be relinquished, even though it may cause a hardship to particular individuals.

Plaintiffs challenged the homosexual exclusion policy on a wide spectrum of legal claims and theories based on every type of individual right ever established under the United States Constitution. But these challenges, naturally, focused on the individual, the human element, while ignoring the larger implications of the policy— as in *Beller* where "[d]uring the discharge hearings of the plaintiffs, various members who testified on their behalf indicated that *while plaintiffs' homosexuality did not impair* the efficiency of the Navy, a member's homosexual conduct might *in other circumstances cause difficulties, especially aboard a ship*."[42] Thus, this human element—the adverse impact of the homosexual exclusion policy on *individual homosexuals*—rather than some aspect of the broad policy issues involved, was the essence of plaintiffs' arguments that the policy was so irrational it violated equal protection guarantees.

Clearly, however, the effect on certain individuals is not the determinant of a policy's rationality or even appropriateness. From the viewpoint of managing an army, a challenge to a military personnel policy is not a hermetically sealed question that affects only the individual plaintiff, and thus should be decided on individual equities. The Army must look at its policies *ex ante*. It must assess the cumulative cost to the Army of providing *all* homosexuals a *right*— in economic terms, an *"incentive"*—to seek military service.[43] By contrast, plaintiffs' *ex post* arguments—that is, arguments based on the effect of the exclusion policy on individuals—simply do not respond to broad policy concerns.[44]

Judge Easterbrook, presently sitting on the Court of Appeals for the Seventh Circuit, has keenly articulated the difference between an *ex post* and *ex ante* analysis of legal questions, and his explanation of these analytical approaches sheds a strong light on the perspectives of the parties in a challenge to the homosexual exclusion policy. Judge Easterbrook illustrated the difference between the *ex post* and *ex ante* analytical approaches with Clark v. Community for Creative Non-Violence (*CCNV*).[45]

CCNV involved a National Park Service regulation that proscribed camping in parks in Washington, D.C. The Park Service, however, did allow the erection of "symbolic" tents in parks near the White House as part of a demonstration for the homeless. Once the tents were up, the demonstrators asked permission to sleep in the tents. The Park Service refused. The court of appeals, citing the First Amendment, found for the demonstrators. Judge Easterbrook observed:

The points made by the court of appeals [in *CCNV*] are typical of ex post arguments in litigation. They start from the assumption that the parties occupy fixed positions and ask whether the distribution of entitlements *in that position* is fair. Here the court . . . started from the existence of a village of tents occupied by the CCNV and asked: Why not allow a little sleep? The effect is small, the demonstrators few, the need great; less restrictive alternatives are available. Only knaves would say no.

Yet the Supreme Court said no. It simply changed the perspective. Instead of asking, "What is the effect of allowing these few people to close their eyes?", it asked, in essence, "What is the effect of camping in the national parks near the White House?" . . . [T]he Court asked not would happen if the CCNV's demonstrators took naps, but what would happen if a *relaxation of the ban on camping made similar demonstrations more attractive.* As the implicit cost of demonstrating fell, more people with less to say would seek initial permits; those who sought permits would stay longer; those whose desire to speak was weaker than the CCNV's would come more frequently; some impostors who simply wanted to sleep and not speak would use the parks as living quarters. . . . The Court was barely interested in the CCNV's methods and message. It asked instead *how complex patterns of behavior would change* if sleeping were permitted.[46]

The Supreme Court's approach in *CCNV* was *ex ante*—that is, forward-looking, considering effects on the system as a whole. Like the plaintiff in *CCNV*, plaintiffs challenging the homosexual exclusion policy asked, "What is the effect of allowing me—*one individual homosexual*—to be in the military?" The proper question, however, is, "what is the effect of allowing *all* homosexuals to be in the military?" What would happen if a *relaxation of the ban on homosexuality made military service more attractive to homosexuals?* How would *complex patterns of behavior change* if homosexuality were permitted within the military?

These proper and important questions are addressed in detail below. Nevertheless, because plaintiffs' arguments regarding the effect of the homosexual policy on them as individuals were the subject of wide discussion, they will be briefly examined.

Plaintiffs' arguments can be summed up and expressed in three general points. *First*, the homosexual exclusion policy is irrational because it punishes homosexuals for something—their homosexuality—they did not choose and cannot change. *Second*, the homosexual exclusion policy is irrational because it ignores individual equities. And *third*, the homosexual exclusion policy is irrational because it wrongly emphasizes or relies upon social and political dynamics as they *are*, rather than as they *should be*. These points will be discussed in turn.

PUNISHMENT

Plaintiffs argued the homosexual exclusion policy is irrational because it punishes homosexuals for their homosexuality, which is a trait they did not choose and cannot change.[47] In one judge's view, for example,

When the government discriminates against homosexuals [or single parents, overweight persons, transsexuals, the young, or the old, for example], it is discriminating against persons because of what they are, through *no choice* of their own, and what they are [or may be] *unable* to change. Preventing such unfair discrimination is what the equal protection clause is all about.[48]

Even if choice—that is, personal responsibility or, as some denote it, "fault"—is relevant to constitutional analysis, it is by no means clear that homosexuals could establish they did not choose their orientation and they cannot change it. Although experts certainly disagree, there have been reports that about one-third or more of those wishing to change their homosexual orientation became exclusively heterosexual after psychoanalysis or psychotherapy.[49] Nevertheless, this idea that excluding homosexuals from the military punishes them for something they did not choose and cannot change is attractive at first blush. It does not, however, withstand close analysis.

One defect in this argument is that not every result unsatisfactory to the individual is recognizable as "punishment," either at law or in logic. "Punishment," as contemplated by the law, quintessentially is the result of criminal sanctions. Administrative results simply do not impose "punishment."

The logical conclusion, then, is that when a military personnel policy operates to exclude an individual, "[n]o punishment is involved . . . [because the] Secretary has the statutory authority to decide that certain attributes make one ineligible for military service."[50] This proposition is true even when exclusion is ancillary to judicially-imposed punishment, as when a soldier is discharged from the military based on a civilian conviction. In that case, courts have held, the challenged regulation "does not impose additional punishment, but merely provides the Army with a means of removing from its ranks undesirable officers who have committed serious crimes."[51] Moreover, even if homosexual conduct had some measure of constitutional protection, "it does not follow that the Army could not exclude homosexuals."[52] Plainly, the concept of "punishment"—in legal, logical, and practical terms—is simply out of place in any discussion of the homosexual exclusion policy.

The "punishment argument" ultimately fails, however, because of the logical defect in maintaining denial of the opportunity to become a soldier impacts at all on one's ability to be a homosexual. In *Dronenburg*, plaintiff argued:

the government should not interfere with an individual's freedom to control intimate personal decisions regarding his or her own body, except by the least restrictive means available and in the presence of a compelling state interest.[53]

If the homosexual exclusion policy interfered with a person's "freedom to control intimate personal decisions regarding his or her own body," this argument might have some commerce. As it is, however, the homosexual exclusion policy controls *accession to the armed forces.* It does not control an individual's sexual activities or intimate personal decisions. The policy does not operate differently from any other statutory scheme. It no more interferes with an individual's ability to pursue homosexuality than excluding certain individuals from a welfare entitlement scheme "interfere[s] with a family's fundamental right to live in the type of family unit it chooses."[54]

Even if the operation of the homosexual exclusion policy somehow did punish homosexuals, the premise that punishment is based on something homosexuals did not choose and cannot change wholly misses the point. Much judicial and academic time has been invested in exploring the etiology of homosexuality.[55]

In *Baker*, for example, the district court heard abundant expert testimony and then found,

Although there are different theories about the "cause" of homosexuality, the overwhelming majority of experts agree that individuals become homosexuals because of biological or genetic factors, or environmental conditioning, or a combination of these and other causes.[56]

On close review, the court's finding encompasses all possible causes of sexual behavior, including *learned behavior,* which is an aspect of environmental conditioning. Thus, even if etiology were relevant, it is not clear that the cause of homosexuality has been conclusively established so as to be useful in legal or practical analysis. This debate on how or why individuals become homosexual no doubt will rage into the foreseeable future. Fortunately for policy-makers, this debate need not concern them *because how or why individuals become homosexual is completely irrelevant to the secre-*

tarial determination that homosexuality is incompatible with military service.

Courts have reached this same conclusion in cases besides those challenging the military's homosexual exclusion policy. In *Aumiller*, a case involving a homosexual professor challenging his termination from a university, the court heard expert testimony on the etiology of homosexuality and its psychological implications, but questioned the relevance of such evidence to plaintiff's First Amendment speech and association claims.[57] In a case involving issues of "whether and to what extent the mental illness is an 'immutable characteristic determined solely by the accident of birth,'" the court declined to base its ruling on the etiology of mental illness.[58]

Nevertheless, some urge that the etiology of homosexuality is relevant to "fault."[59] As one judge urged in reviewing a security clearance procedure based on homosexuality, the determinative factor for deciding the constitutionality of behavior-based regulations was "what causes the behavior? Does it arise from the kind of characteristic that belongs peculiarly to a group that the equal protection clause *should specially protect*?"[60] This type of causative focus is sometimes seen in sociology and medicine as well. For example, in an article titled *AIDS and Compassion*, Dr. Friedland writes:

> Rather than blaming the victim, we should examine the societal conditions that promote unhealthy risk-taking behavior, particularly intravenous drug use, and ask who is responsible for them? . . . Illicit drug use is entwined with and the product of unemployment, poverty, racism, and hopelessness and is perpetuated by greed and corruption at many levels in our society. . . . Certainly the drug user is responsible for his or her own actions, but he or she is also the victim of powerful and shameful social forces and conditions.[61]

Similarly, the "blame" for many service-disqualifiers—for example, single parenthood or failure to graduate from high school—could be apportioned between the individual and society. The fact that social forces may play a role in creating an individual's circumstances, however, is not relevant to whether or not those circumstances are compatible with military service.

Thus, *fault*—however that term is used, either in regard to homosexuality or to any other service-disqualifier—simply is not relevant to military personnel policies. The Army does not attempt to determine the etiology of single parenthood, drug or alcohol abuse, high school absenteeism, obesity, low mental acuity, or of any other service-disqualifying characteristic. When a person is disqualified from military service by reason of physical or mental handicap, for example, whether he was handicapped by accident of nature or through his own fault is irrelevant wholly to the fact he does not meet enlistment or retention standards for soldiering.

The Army must make personnel decisions and apply broad policy objectives based on an individual's situation at the time he presents himself for enlistment, not on some notion of whether the individual was responsible or not for creating that situation. The Army plainly is not in the business of sorting out the genesis of social or sexual phenomena. Nor should the Army be expected to attempt that formidable task. Finally, even assuming the homosexual exclusion policy punishes homosexuals for something they did not choose and cannot change, this fact would not demonstrate that homosexuality *is* compatible with military service.[62] As the *Robinson* court noted, for example, "a person may even be a narcotics addict from the moment of his birth."[63] Clearly, that fact—that an addictive personality may be unchosen and unchangeable—does nothing to demonstrate addiction is compatible with military service. No different conclusion is warranted in regard to homosexuality.

The Army simply sets accession and retention standards for soldiers. These standards are based on the needs of the service, which vary and fluctuate over time. Such standards are not designed—nor do they operate—to "punish" anyone. This proposition is true regardless of how a particular individual acquired the characteristic that ultimately may disqualify him from military service, and regardless of whether the individual deplores the service-disqualifying characteristic or prefers it. As Chief Justice Burger observed in *Hardwick*, the regulation of homosexual conduct "is essentially not a question of personal 'preferences' but rather of the legislative authority of the State."[64] Similarly, the homosexual exclusion policy is not about personal sexual preferences. Rather, the policy is one expression of the judgment of senior military and civilian leaders on how to maintain an "efficient and easily administered system of raising armies"[65] whose "primary business . . . [is] to fight or be ready to fight wars should the occasion arise."[66]

Case-by-Case Analysis: The Individual Equities

The second argument advanced by plaintiffs in an attempt to show the homosexual exclusion policy is irrational asserts that the policy produces unfair, even foolish results in the case of particular homosexuals. Plaintiffs claim the policy can be rational only if it operates on a case-by-case basis and allows for individual equities and due regard for the "distinguishing facts of plaintiff's case."[67] The Supreme Court repeatedly has recognized that individual preference is not one of the criteria for constitutional soundness in policy-making. The Court has stated, "[a]s long as the classification scheme . . . rationally advances a reasonable and identifiable governmental objective, we must disregard the existence of other methods of [operation] that we, as individuals, perhaps would have preferred."[68] Plaintiffs' argument for case-by-case evaluation of the impact of *their* homosexuality on the military again highlights the difference between *ex post* arguments, which are not responsive to broad policy concerns, and *ex ante* analysis, which is.

Plaintiffs' claims are based, essentially, on their individual military service records. This reliance on service records is somewhat ironic considering many plaintiffs entered military service by evading the homosexual exclusion policy in the first instance. Thus, no legitimate basis for compiling a service record existed.[69] Even so, from the Army's perspective, the service record of the individual plaintiff simply is not material to whether or not homosexuality—as a set of discernible characteristics of a class—is compatible with military service.

Policy-makers know fully well that *all* military exclusion policies sometimes exclude *good soldiers*. Yet that result, like the result of the exclusionary rule for evidence obtained in violation of the Fourth Amendment, is the considered price military society is willing to pay for good order, discipline, and morale within the armed forces. Thus, a paraphrase of Justice Harlan's famous statement applies here to the judgment calls underlying military exclusion policies: "it is far worse to enlist a poor soldier than to let a good soldier go free."[70] This perspective, unlike plaintiffs' view, recognizes the reality of how personnel policies must operate if they are to be in fact *policies*.

Policies sometimes must rely on classwide presumptions, and policy-makers are entitled to make presumptions knowing fully well not every individual subject to the presumption will fit the presumption. As the Supreme Court made clear in *Gilliard*, "Congress is entitled to rely on a *classwide presumption* that custodial parents have used . . . support funds in a way that is beneficial to entire family units," even though there was evidence some parents did not use funds to benefit the family.[71]

The reality of policy-making is that "[n]early any statute which classifies people may be irrational as applied in particular cases."[72] Further, as the Supreme Court has observed, "the drawing of lines that create distinctions is peculiarly a legislative task and an unavoidable one. . . . Perfection in making the necessary classifications is neither possible nor necessary."[73]

These points are especially apt in the context of military personnel policies. The scope of responsibility in military personnel management is immense, extending even to combat readiness. Not only must military personnel policies deliver soldiers with the best potential for successful service, but they must do so without imposing an undue administrative burden on the Army's mission to fight wars. The courts have recognized that fiscal and administrative burdens are proper factors to consider in determining the constitutionality of policies based on due process challenges.[74] Even greater weight should be given to these factors when the challenge to the policy implicates combat readiness. Indeed, no soldier who has ever served has not at some time wished his personal situation could receive more individualized attention from the Army. In the end, however, the Army determines—as it must—when individualized personnel decisions profit the Army as a whole. In regard to the homosexual exclusion policy, the *Beller* court observed that the military "could

rationally conclude that homosexuality presented problems sufficiently serious to justify a policy of mandatory discharge while other
grounds for discharge did not."[75]

The law clearly does not require the Army to "show with partic-
ularity that the reasons for the general policy of discharging homosexuals . . . exist in a particular case before discharge is permitted."[76] Furthermore, the "[d]ischarge of the particular plaintiffs . . .
[is] rational, under minimal scrutiny, not because their particular
cases present the dangers which justify [the homosexual exclusion]
policy, but because the general policy of discharging all homosexuals is rational."[77]

The argument urged by plaintiffs, however, is that the Army
should make a showing as to why a particular homosexual should be
discharged. This particularized showing, plaintiffs contend, would
result in a workable, better balance of the homosexual's desire for
military service and the needs of the Army.

Plaintiffs ask the Army to take on a huge administrative burden
in a futile attempt to predict the unpredictable. In a case brought by
a homosexual who had been denied a security clearance, the court
observed, "[the] attempt to define not only the individual's future
actions, but those of outside and unknown influences, renders the
'grant or denial of security clearances . . . an inexact science at
best.'"[78] Even the phrase "inexact science" is too generous to denote
the proposed process of predicting which homosexuals would violate the Uniform Code of Military Justice's proscription of sodomy,
engage in service-discrediting homosexual conduct, or otherwise
disrupt the ranks, and which would not. *No* measure or method—
much less any analytical framework resembling "science"—exists
by which to make this prediction.

Indeed, in cases challenging the homosexual exclusion policy,
plaintiffs frankly refused to venture a prediction of their own future
conduct.[79] To the contrary, in *Steffan*, the plaintiff sought a protective order and invoked his Fifth Amendment privilege to avoid
answering discovery questions concerning homosexual conduct.[80]

This lack of candor obviously has complex social and personal
dimensions. Moreover, so long as sodomy remains a crime or is
subject to social opprobrium, and so long as considerations relevant

to the situation of sexual partners exist, candor will continue to be diminished. Even medical researchers have found that not all people readily discuss their sexual history with health care providers. For example, "[s]tudies have shown that some men who claim they caught AIDS from a prostitute . . . eventually admit to having practiced such risky behavior as intravenous drug use or homosexual anal intercourse."[81] It is even more unlikely that most individuals would participate in a detailed, candid discussion with Army recruiters about their past, present, or hoped-for future sex lives.

Thus, it is fair to say individuals have different levels of tolerance for personal sexual candor. Even if the Army could somehow sort out which homosexuals would involve themselves in homosexual conduct or other disruptive behavior and which would not, any lack of candor would quickly defeat the system and the Army would again be in the position of being unable to predict the future.

The fact is, no one—not even the individuals involved—can say with certainty how their homosexuality will, over time, affect their service as a soldier. As one ex-soldier explained, he eventually became "very [homosexually] active with a number of soldiers on the post. I had been there almost five years and my tour was almost up, so I began throwing caution to the wind. I wasn't much into the psychological mating or lover relationship."[82]

Human nature would break the mold if most homosexuals—like other people—did not, at least eventually, give in to their sexual desires. One homosexual ex-soldier told how her good intentions failed:

> I entered the military *knowing* that I was a lesbian, but also knowing that I wanted to do what was right by military standards and stay there! But, by God, when I got into basic, I thought I had been transferred to hog heaven! No damn kidding! Lordy![83]

In the event criminal or disruptive homosexual behavior does result, the Army then must bear the additional burden and considerable expense of administrative or judicial discharge of the individual. Of the 42 homosexuals whose interviews appear in Humphrey's book, *My Country, My Right to Serve*, which covers the period from 1940 to 1990, 31 individuals reminisced about involvement in at least one investigation or administrative or judicial proceeding based on homosexuality during their time in the military.[84]

Moreover, when a policy allows for exception or individualized determinations, this serves to provide another basis for litigation.[85] In *Berg*, for example, the court stated, "[a]lthough the Navy regulation on homosexuality . . . does not in terms provide any exception to the general policy of separating homosexuals the Navy has interpreted it as not mandating separation in all cases."[86]

The *Berg* court then found the Navy had failed to "articulate adequately why it determined not to retain this appellant."[87] In practical terms, the signal this sends to the agency is that a policy that allows for—or in fact operates on—a case-by-case analysis will be subjected to a higher standard of proof (in essence, a higher standard of review) than an across-the-board policy subjected to the rational basis test. In Doe v. Casey, the court observed,

> Because Doe himself does not view homosexuality as stigmatizing . . .
> he would have no liberty interest claim if *all* homosexuals were banned
> from CIA employment. If, on the other hand, the CIA terminated Doe
> because his homosexuality presented a *unique security risk* not neces-
> sarily presented by all other homosexuals, we must conclude that the
> statement is sufficiently stigmatizing to give rise to a colorable liberty
> interest claim.[88]

Thus, if one policy goal is to avoid the drain of litigation—and this is a laudable goal for a number of military and social reasons—avoiding policies that provide exceptions or individualized determinations—except where constitutionally required—may help accomplish that goal. Moreover, unlike some other grounds for discharge where the Army has decided to provide for exceptions or case-by-case determinations, the Army reasonably could conclude that to provide exceptions as to homosexuality would compel the Army "to engage in sleuthing of soldiers' personal relationships for evidence of homosexual conduct in order to enforce its ban on homosexual acts."[89]

This point was addressed explicitly in *Beller* in response to plaintiff's argument that homosexuals should not be subject to automatic discharge when some other individuals were subject to discharge, but discharge was not automatic. The *Beller* court stated,

[T]he district court . . . appeared to declare due process violated because some other groups subject to discharge were not required to be discharged. . . . Some personnel are given a second chance to "overcome his/her deficiencies". . . . Giving someone a second chance to overcome his or her deficiencies is not at all the same as requiring fitness of the individual to be considered. In any event, the fact that the [military's] choice of categorization is overinclusive and underinclusive does not mean that the regulations violate due process. . . . The [military] could rationally conclude that homosexuality presented problems sufficiently serious to justify a policy of mandatory discharge while other grounds for discharge did not.[90]

Drug use, for example—which may be waived as a service-disqualifier—can be detected and deterred by urinalysis, which is hardly an elaborate or intrusive investigative method. Homosexual acts, by contrast, like other sexual misconduct, are difficult to prove and more difficult to deter, even by imposition of criminal sanctions or the threat of fatal disease.

The Army need not choose between sleuthing and "[shutting] its eyes to the practical realities" of homosexuals in the military.[91] These practical realities must be considered both for the Army and the individual. Several homosexuals who evaded the policy and served in the military in spite of it have described their experiences, and—by implication—the experience of the military with homosexual service members. One former soldier, who eventually was convicted at court-martial for committing sodomy with an enlisted man in the barracks, told:

I was ordered in front of the post commander. He said, "We have an accusation that you're a homosexual. It's being investigated by [Criminal Investigation Division], and you are not to go to your work station at the vault [a classified area] until we find out what's going on." I was quaking in my boots because *it was true, it was quite true that I was gay,* but I thought nobody knew. . . . However, *nothing was found in the investigation, so he later called me in and apologized.*"[92]

A former sailor gave this description of some of his experiences as a homosexual in the Navy:

I was called in by the [Naval Investigative Service]. . . . Apparently, they had been observing me for a while and had signed affidavits from different people that stated I had engaged in homosexual acts. At that time, *I denied it, of course,* but with what they had, the commanding officer recommended a court-martial. . . . Then I got arrested in Milwaukee, by civilian police, for indecent exposure. It was like all these things were coming down at one time. . . . I was under quite a bit of pressure.[93]

Finally, a former Air Force officer and squadron commander, gave this account of how he ran afoul of the homosexual exclusion policy:

We [the male captain and an enlisted male airman] were stroking, caressing, and kissing in front of the other party guests . . . One of the other guys at the party was in my squadron [I was the Commanding Officer of the squadron]. . . . This [Air Force Office of Special Investigations] investigator, a woman, said, "Would you like to make a statement?" . . . I wrote, *"I am not now nor have I ever been homosexual, nor have I committed any homosexual acts,"* signed Paul Starr. *That* came back to haunt me later on in the court-martial for making a false official statement. That was one of my charges, but at that time I did not know that they were putting my case together from many sources.[94]

Plaintiff Hatheway, for example, detailed his homosexual activities while in the Army in an interview after his discharge.[95] At his trial by court-martial for sodomy, Hatheway "raised a defense of unconsciousness. One psychiatrist testified Hatheway suffered from 'pathological intoxication' and stated his opinion that Hatheway was not conscious of the sodomitic acts. Another psychiatrist testified this was a real possibility."[96]

Every defendant has the right to put the government to its proof on a criminal charge. But, in making policy decisions, the Army properly can consider the administrative and judicial burdens— even a forced requirement to sleuth—presented by homosexual conduct. It should also be noted that, in military practice, defense experts—such as the psychiatrists who testified for Hatheway—generally are paid for by the military, not the defendant.[97] Clearly, a policy-maker rationally could conclude that a policy which avoids or minimizes a choice between sleuthing and "[shutting] its eyes to the practical realities" of homosexuality in the military in the first instance is best for the Army and the individuals involved.

The homosexual exclusion policy sets a general rule but, as the Supreme Court has stated in another context, "[g]eneral rules are essential if [an army] of this magnitude is to be administered with a modicum of efficiency."[98] This is true "even though such rules inevitably produce seemingly arbitrary consequences in some individual cases."[99] Thus, when plaintiffs argue that individual equities demonstrate the homosexual exclusion policy is irrational, they simply misperceive the focus and operation of military personnel policies.

What Should Be, According to Plaintiffs

The final major theme disputing the rational basis of the homosexual exclusion policy centers on the notion that the policy wrongly emphasizes or relies upon social and political dynamics as they *are*, rather than as they *should be, according to plaintiffs*. This theme has three notable variations. These are first, that the homosexual exclusion policy is irrational because it is based on "morality," which should have nothing to do with the law; second, that the policy is irrational because it is based on societal prejudice against homosexuality, a prejudice the Army should endeavor to change; and finally, that the homosexual exclusion policy is irrational because it reaches the off-duty conduct of soldiers—conduct that should not be the Army's proper concern. The next three sections will discuss these arguments in turn.

Morality

Plaintiffs challenging the homosexual exclusion policy have often
disputed the moral dynamic of homosexuality. Some have claimed
the policy is based on morality and therefore the policy abridges the
Establishment Clause.[100] More boldly—because the law always has
recognized that it "should . . . be flexible enough to recognize the
moral dimension of man and his instincts concerning that which is
honorable, decent, and right"[101]—some plaintiffs have contended
that "the existence of moral disapproval for certain types of behavior
is the very fact that disables the government from regulating it."[102]
Indeed, one judge concluded: "[It is] of *no significance* that the
[homosexual] conduct in question may be condemned as immoral
by a majority in our society."[103] As in *Hardwick*, plaintiffs insisted
"that majority sentiments about the morality of homosexuality
should be declared inadequate."[104] The argument, then, has often
been that the homosexual exclusion policy is irrational because it
coincides with "majority morality and majority choice is always
made presumptively invalid by the Constitution."[105]

Courts readily found that this theory of irrationality is not
grounded in law or logic.[106] The Supreme Court has made clear—
expressly so as to homosexuality—that morality informs the law.[107]
In *Hardwick*, the Court stated,

> [Morality] is said to be an inadequate rationale to support the law
> [proscribing sodomy]. The law, however, is constantly based on no-
> tions of morality, and if all laws representing essentially moral choices
> are to be invalidated under the Due Process Clause, the courts will be
> very busy indeed.[108]

Though moral concerns clearly are sufficient to meet the rational
basis test, the military interests furthered by the homosexual exclusion policy are much broader than the suppression of moral delin-
quency.[109] Still, it is true that there are a variety of views on the
moral dynamics of homosexuality. The fact the homosexual exclusion policy may coincide with an historical, moral, or majoritarian
view of homosexuality, however, does not mean it lacks a rational

basis.[110] Indeed, one can only imagine the type of army that might be fielded if—to be rational under the rational basis test—military personnel policies were required to take an approach that was the opposite of historical, moral, and majoritarian views.

Finally, it is important to recognize that morality is not the law's poor cousin: on the contrary, "[i]n military life there is a higher code termed honor, which holds its society to strict accountability; and it is not desirable that the standard of the Army shall come down to the requirements of a criminal code."[111] This "higher code termed honor" is much more a part of every day military life than might be apparent to those not in uniform. This "higher code," for example, is always applied in addition to the requirements of the military's criminal code.[112] Moreover, moral precepts have been an important, explicit, and integral part of soldiering from the time of the very formation of the United States Army.

Prejudice

A second theory of irrationality centers on prejudice. Plaintiffs claim the homosexual exclusion policy lacks a rational basis because it accounts for, or is based on, existing societal conclusions or attitudes about homosexuality. A corollary to this theory was that the pertinent societal conclusions or attitudes about homosexuality themselves had no basis in fact.

Under this theory of "prejudice," the Army's justifications for excluding homosexuals amount to no more than a "series of platitudes."[113] The policy, plaintiffs contend, only serves to "illegitimately cater to private biases."[114] Moreover, in this view, the Army should—or should be forced to—take up the task of endeavoring to change societal attitudes or prejudice toward homosexuality.

This task of social reformation—as urged by plaintiffs—takes on huge proportions. It extends even to changing the attitudes of foreign nations and secret intelligence organizations. In *High Tech Gays*, where homosexuals employed by defense contractors challenged a security clearance procedure, plaintiffs argued the KGB's doctrine—shown to include targeting homosexuals—should be ignored in favor of applying the position of the American Psychological Association on homosexuality to issues in the case.[115] In Miller v. Rumsfeld, a judge endorsed the argument that combat readiness was unrelated to—or at least unaffected by—how citizens of host nations viewed the conduct of American military members.[116]

Miller notwithstanding, the way in which a host nation views the United States Armed Forces is critical indeed. In *Beller*, the court held that the Navy reasonably could deem that the homosexual exclusion policy ensures "the acceptance of men and women in the military, who are sometimes stationed in foreign countries with cultures different from our own."[117] Indeed, this very issue was faced in the Gulf War. A Pentagon spokesman noted:

[H]omosexual acts are punishable in Saudi Arabia by execution. When service members assigned to Saudi Arabia during Operation Desert Shield and Desert Storm were caught in homosexual acts, they were returned immediately to the United States and discharged administratively. . . . Immediate evacuation avoided having the incidents come to the attention of Saudi religious police.[118]

In United States Information Agency v. Krc, a government employee was properly denied overseas assignments because of homosexuality.[119] Likewise in Padula v. Webster, the FBI, as a national law enforcement agency, was entitled to conclude that agents who engage in homosexual conduct criminal in roughly half the states would undermine the agency's effectiveness.[120] Finally, in Sullivan v. Immigration and Naturalization Service, a homosexual claimed on his own behalf that "deportation to Australia will cause him undue hardship because homosexuals are not accepted in that society."[121] Taking account of various social realities, whether of nations or among spymasters, is not countenancing prejudice. It is using common sense to compose a fighting force that can operate anywhere in the world without provoking unnecessary social controversy or opposition.

In challenges to the homosexual exclusion policy, plaintiffs deftly equate social controversy or opposition to homosexuality with "prejudice" toward homosexuals. This equation is sophism. The argument that the homosexual exclusion policy has its genesis in social prejudice fails, first, to admit that the homosexual exclusion policy is premised on the risk of eventual homosexual conduct. The court in *Ben-Shalom* easily made this distinction by finding that the homosexual exclusion policy "does not classify plaintiff based merely upon her status as a lesbian, but upon reasonable inferences about her probable conduct in the past and in the future."[122]

The argument that the homosexual exclusion policy is an expression of prejudice toward homosexuals assumes that attitudes about homosexuals exist in the abstract and are unrelated to attitudes about homosexual conduct. It was in this context that plaintiffs often analogized the homosexual exclusion policy to racial segregation.

The analogy between homosexuality and race, however, compares two incomparable traits. Even some courts positing this analogy conceded it was flawed by the fact that "constitutionally, race is a suspect classification; and race, unlike desire or intent, cannot be suppressed."[123] On the other hand, some courts accepted the analogy without discerning its flaw, as in Saal v. Middendorf, where the court stated the military,

> has abandoned the *stereotypes* of the past that have stigmatized women and members of minority races. . . . [I]ts failure to accord the same treatment to plaintiff *simply because she engaged in homosexual acts* must be found to be irrational and capricious and thus in violation of the Fifth Amendment.[124]

The *Saal* court breezily compared two incomparable matters: stereotypes—that is, *attitudes*—and *conduct*. Nevertheless, the analogy between homosexuality and race frequently is held up as grounds for skepticism concerning the homosexual exclusion policy.[125]

Such skepticism—the nagging idea that excluding homosexuals from the military is like segregating blacks into separate battalions—is yet another expression of the status-conduct dichotomy, which is the dividing line of analysis not only in homosexuals' challenge to the exclusion policy but in all questions of extending "rights" or privileges to behavior-based groups. One commentator ably explained the respective sides in an on-going debate on a city sexual preference law as follows: "[c]ritics of homosexuals project a city of immorality, filled with gay *sex* in public restrooms [*conduct*] . . . while the gay rights supporters speak in broad, impassioned terms about *equality* [for their homosexual *status*]."[126] Because homosexual conduct does not have constitutional protection and, indeed, frequently is against the law, no theory of prejudice can survive without a strict application of the status-conduct dichotomy.

Thus, plaintiffs challenging the homosexual exclusion policy claimed—apart from any prospect of conduct all parties agreed was detrimental to the Army—that homosexuals are excluded from military service "on the ground that they [personally and apart from homosexual conduct] are offensive to the majority or to the military's view of what is socially acceptable."[127]

Joining the analogy to racial segregation, plaintiffs argued that if homosexuals properly could be excluded from military service, then "no rights are safe from encroachment and no minority is protected against discrimination."[128] Thus, plaintiffs attempted to demonstrate the homosexual exclusion policy was *irrational* because racial segregation was *impermissible*.

One court flatly declared that the argument linking homosexuals and racial minorities was "completely frivolous."[129] Conversely, one court rejected the Army's position as an attempt "to avoid the race analogy by arguing that the homosexuality regulation is directed at conduct," whereas past policies on segregation were based upon attitudes among the races.[130]

The Army's position—that racial segregation was based on attitudes among the races while the homosexual exclusion policy was directed at homosexual conduct—is borne out by the military's response to racial integration. The military, to the present, has an extensive equal opportunity program designed to shape and educate attitudes among the races. The equal opportunity program does not focus on behaviors because there are no behaviors that specifically characterize a particular race.

If a similar program were implemented on behalf of homosexuals, however, it could not succeed in shaping or changing attitudes toward homosexuals unless it *shaped or changed attitudes toward homosexual conduct*. Such a program could change attitudes toward homosexuals only if it taught people: (1) that homosexuals do not, or rarely, commit homosexual acts, or (2) that homosexual acts are morally and socially acceptable, even if legally proscribed. The first approach is not demonstrably true, and the second approach is decidedly outside the Army's charter.

Moreover, the jurisprudential ramifications—and hence the social results—of such a program should be considered. One observer predicts "[g]ranting special protection for any *behavior-based* life style will eventually undermine the very core of the judicial system by removing its ability to *evaluate a person's character based on his or her conduct.*"[131] Thus, this flawed analogy, in the end, failed to gain constitutional stature.

The argument equating homosexuals with racial minorities has been pursued and rejected outside the courtroom as well. A long-time civil rights activist observed:

> By positioning themselves as a minority, homosexuals skillfully used the rhetoric of civil rights ("discrimination") and the widespread public legitimacy of the civil rights movement to advance their political goals.

> As a reverend who endeavored to advance the civil rights movement in conservative areas . . . , I have found it increasingly curious that many people uncritically, indeed reflexively, accept this comparison of homosexuals to minorities. To equate a status fixed by genetics (sex, race) or historical accident (national origin) with a behavior-based status (a group linked by a common preference, say, to commit sodomy) requires serious analysis.

> . . .

> The only answer is the mesmerizing impact of a misplaced analogy. The person who finds sexual expression in sodomy (or, for that matter, in incest, adultery, bestiality or sadomasochism) can define himself as a member of a particular "minority" who shares that behavior. But it makes no sense for him to claim the same social protections of those who are born black, female or in a foreign country.[132]

One homosexual activist admits, "Black people tell me, 'Don't compare what you're doing [for homosexual rights] to what civil rights leaders of the '60s did,' . . . [b]ut I don't see any difference."[133]

Those who do not see the difference between homosexual rights and civil rights might look closely at Dr. Martin Luther King, Jr.'s poignant statement. He said, "I have a dream my four little children will one day live in a nation where they will not be judged by *the color of their skin* but by the content of their character."[134] Dr. King's dream was the opposite of the status-conduct dichotomy. His dream was that Afro-Americans *would be judged by their conduct*—that is, the outward manifestation of their individual character, beliefs, ideals, values—rather than by a "status fixed by genetics . . . or historical accident"—that is, the color of their skin.

Thus, the analogy between those who commit homosexual acts and those who have a certain skin color has no utility. Moreover, regardless of the usefulness of the analogy, the comparison between homosexuality and race does not address—much less negate—the bases for the homosexual exclusion policy.

Racial segregation in the military was a sad expression of a larger segregation based on some "view of what [was] socially acceptable" at the time. As will be discussed, however, the concerns underlying the homosexual exclusion policy are infinitely broader than social sensibilities.

Homosexuality, unlike race, affects military security and privacy concerns, as well as assignment and deployment considerations. A member of a racial minority is not subject to blackmail on pain of revealing his minority membership. Homosexuality, however, is well recognized as a basis for blackmail.[135] Further, race does not implicate privacy entitlements because privacy entitlements are related to gender. Privacy entitlements are implicated by homosexuality, however, because—for homosexuals—gender segregation does not achieve segregation by sexual preference.[136]

Homosexual culture, to the extent that it involves homosexual conduct, cannot be integrated into the military community in the same way as racial minority cultures are integrated. Cultural expressions of racial and homosexual cultures do not present similar legal, social, political, or moral issues. For example, no commander would hesitate to authorize the post exchange to sell *Ebony* or *Jet*, magazines oriented to black culture. Radically different issues and problems would be presented by sale at the post exchange of *The*

Advocate, the nation's largest publication oriented to homosexuals.[137]

The homosexual exclusion policy has a moral dimension, one that did not attend racial segregation. Finally, because homosexuality—unlike racial membership—does not have constitutional status, a policy integrating homosexuals into the military would require *creating*—rather than *implementing*—rights for homosexuals. Creating rights is a vastly different, and potentially more disruptive, social process than implementing the law of the land.

Racial segregation and homosexual exclusion clearly are not alike, either in legal status or in practical terms. Simply put, racial minorities never have and never will present the potential that homosexuals present for adversely affecting military interests on several fronts.[138] Thus, the analogy between homosexuality and race provides no basis to find the homosexual exclusion policy irrational or to claim the rights of all minorities are in jeopardy. As the court carefully explained in *Dronenburg,*

> The Constitution has provisions that create specific rights. These protect, among others, racial, ethnic, and religious minorities. If a court refused to create a new constitutional right to protect homosexual conduct, the court does not thereby destroy established constitutional rights that are solidly based in constitutional text and history.[139]

The analogy between homosexuality and race quickly breaks down. Analogies, nevertheless, are helpful in revealing the assumption underlying plaintiffs' claims that homosexuals are excluded from military service solely as a result of illegitimate social prejudice. An apt—and hence more illuminating—analogy, however, is one that compares homosexual preference with other sexual preferences, rather than with racial or ethnic identities.

It would be a fundamental analytical mistake to accept the assumption there are only two sexual preferences extant among humans. This assumption, however—like the mesmerizing effect of the equation of homosexuals with racial minorities—often passes uncritically into the analysis of plaintiffs and courts alike. One court took the assumption at face value and found:

We cannot take seriously the dissent's suggestion that the Navy may be constitutionally required to treat heterosexual conduct and homosexual conduct as either morally equivalent or as posing equal dangers to the Navy's mission. Relativism in these matters may or may not be an arguable moral stance . . . but moral relativism is hardly a constitutional command, nor is it, we are certain, the moral stance of a large majority of naval personnel.[140]

This assumption—that only two sexual orientations exist— underlies the notion that if the law would only treat homosexuality and heterosexuality and their respective manifestations as "morally equivalent,"[141] then every person's "core . . . personality, self-image, and sexual identity"[142] would be accommodated and prejudice, as defined by plaintiffs, would be eliminated.

The facts are otherwise. Sexual preferences, whether defined by the partner or object of sexual desire, or by a preference for particular methods of sexual gratification, range over a spectrum limited only by one's imagination. Non-homosexual preferences, such as pedophilia, exhibitionism, fetishism, bestiality, transsexualism, and sado-masochism, are "sexual orientations" in the same sense this term was used by plaintiffs in challenges to the homosexual exclusion policy. In *Dronenburg*, for example, plaintiff argued bestiality could be prohibited, but only on "the ground of cruelty to animals."[143] In *Dronenburg*, plaintiff's view was, but for the fact bestiality was—or may be—cruel to animals, bestiality was a legitimate sexual preference and it could not otherwise be a basis for regulation.

Societal attitudes toward these non-homosexual orientations are based, not on a *prejudgment* of the *individual*, but rather on conclusions about the defining sexual *conduct*. This distinction between having an opinion about an individual and having an opinion about conduct can be seen in societal reaction to AIDS.

The reaction to AIDS generally has been stigmatizing. While some have argued that the stigma attached to AIDS arose from social prejudice against homosexuals as individuals, researchers have observed that social reaction to AIDS is influenced by conclusions about *conduct*.[144] The *Public Health Reporter*, for example, noted

"AIDS is a sensitive issue because it is usually contracted through *behaviors that have negative societal connotations*, such as drug abuse, homosexual behavior, and prostitution."[145] Another researcher noted "negative attitudes about the *lifestyles* of homosexuals and drug abusers are closely bound with attitudes about AIDS."[146] As AIDS becomes less associated with "behaviors that have negative societal connotations," the stigma surrounding AIDS will fade.

This distinction between having an opinion about an individual and having an opinion about conduct may be seen by analogy as well. Sickle cell anemia, for example, which affects primarily members of the black race, imposes no stigma because there is no causal connection between the disease and negatively perceived behavior. Society, on the other hand, has focused on the fact in many cases there is a causal connection between AIDS and conduct. It is unlikely AIDS would carry any stigma if it were caused by living near radio towers—even if more homosexuals than heterosexuals developed AIDS because more homosexuals than heterosexuals lived near radio towers. This is true because living near a radio tower is not a "behavior with negative societal connotations."

Many sexual preferences have negative societal connotations. Unlike the analogy posed between homosexuality and race, however, there is no constitutional, principled way to distinguish between sexual preferences. Apart from the absence of principled, constitutional distinctions, the array of sexual preferences—and their import—frequently presents even a semantic quagmire. One court attempted to distinguish "transsexuals as persons who, unlike homosexuals and transvestites, have sexual identity problems" and found "the term 'sex' does not comprehend 'sexual preference,' but held that it does comprehend 'sexual identity.'"[147] Another court found "a 'homosexual' is one who has an emotional, erotic attachment to one of the same sex—while a 'gay' is one 'who is proud of being homosexual.'"[148] Another court struggling with principled distinctions stated,

the Court will attempt, for the sake of clarity . . . , to use the term "homosexual" in describing specific sexual acts and those persons who engage in, promote, or encourage such acts. The Court will attempt to

use the term "gay" in referring to the more general aspects of the life-
styles of those individuals who prefer the companionship of members
of their own sex and of the commercial institutions that serve those
life-styles.[149]

These semantic troubles would be doubled and tripled if the
courts or the Army were called upon—as they would be if homosexuality obtained constitutional or regulatory protection—to sort out
the "rights" of homosexuals who were also pedophiles, of bisexuals
who occasionally participated in bestiality, of heterosexuals who
gravitated to sado-masochism, and so on into the human sexual
psyche.

There simply is no principled way to distinguish between all the
variations on the spectrum of human sexuality for purposes of
providing constitutional protection to some, but not others. Thus,
even heterosexual expressions are not completely protected by the law. Two
of many examples of this are prohibitions on extramarital relationships[150] and the proscription of prostitution.[151]

The law confines legal protection in sexual matters to "family
relationships, marriage and procreation and does not extend [pro-
tection] more broadly to all kinds of private sexual conduct between
consenting adults."[152] Plaintiffs' claim that the homosexual exclusion policy is an expression of prejudice sets up an "all-or-nothing"
proposition. As *Dronenburg* and other cases recognize, it is "impossible to conclude that a right to homosexual conduct is 'fundamental' or 'implicit in the concept of ordered liberty' unless any and all
private sexual behavior falls within those categories, a conclusion
we are unwilling to draw."[153]

Thus, the body of constitutional law which reserves constitutional protection in sexual matters primarily to family relationships,
marriage, and procreation simply cannot be explained away as "illegitimately cater[ing] to private biases."[154] Nor is illegitimate bias a
reasonable construction of the homosexual exclusion policy.[155] The
bases for the policy are not conjectural[156] and the regulation and its
goals are not nearly "*so remote* as to render the policy arbitrary or
irrational."[157] It simply cannot be said the exclusion of homosexuals
from military service is "*so seldom* relevant to the achievement of

[good order, discipline, and morale in the armed forces] that laws grounded in such considerations can be deemed to reflect prejudice and antipathy," rather than a rational basis in fact.[158]

Off-duty: soldiers or civilians?

The final theory of irrationality strikes at the very nature of military service. This argument asserts that the homosexual exclusion policy is irrational because it is based on the off-duty conduct of soldiers, which—the argument goes—is not the Army's business.

This is not a theme that rings true with the professional soldier: few forget the patient explanation that *they are soldiers twenty-four hours a day*. Indeed, military courts have recognized that "military life is such that the degree of community control over even its law-abiding members is greater than that to which civilian probationers are subject."[159] This unique military status and its import have been affirmed by courts and commentators alike for as long as the Army has existed.[160] This need for discipline stems from the aspect of soldiering that is most unique of all, the fact that while "[n]o man willingly lays down his life for a national cause . . . yet implicit in military life is the concept that he who so serves must be prepared to do so."[161]

In *The Soldier and the Law*, written in 1941 by law professors at the United States Military Academy, the authors explained this unique status in a chapter titled, "To The Soldier":

> Private Doe: "Raise your right hand. Now repeat after me—!" . . . You are now a soldier! Perhaps you are not yet a real soldier in the full meaning of the term—there you stand in your civilian clothes, your civilian haircut, untrained in even the barest fundamentals of military service; you do not *feel* different except for a slight twinge in the pit of your stomach that has nothing to do with food or drink. But a soldier you are, nevertheless. You will quickly attain the outward symbols of military service. The physical development and technical training you will acquire in due time. But *immediately* . . . a change of great moment to you has taken place. You have not merely obtained a new job, a new employer. You are not merely working for the government. You have not merely entered into a contract which you may modify or break at

will. . . . [Y]ou have acquired a new *status*. . . . In short, you have not only entered a new and strange environment . . . , you have become subject *to a new system of law*.[162]

This new system of law quite definitely encompasses the creed, "[t]here is no distinction between duty time and off-duty time as the high moral standards of the service must be maintained at all times."[163] Inside courtrooms where challenges to the homosexual exclusion policy were advanced, however, the impact of military status on the questions before the court was often ignored or slighted. In Matlovich v. Secretary of the Air Force, for example, the court expressed dismay that plaintiff was discharged from the Air Force for *only* "*minimal* [homosexual] sexual involvement with Air Force personnel and none with those with whom he worked."[164]

The idea that the military might be concerned with the off-duty, off-post conduct of its soldiers was often advanced with alarm, if not with outright horror.[165] This horror that a regulation might affect a soldier's off-duty conduct seems surprising when the law clearly countenances the fact that off-duty, off-post homosexual conduct of a soldier is amenable to criminal prosecution under the Uniform Code of Military Justice.[166] Thus, the alarm and dismay that the homosexual exclusion policy countenances off-duty, off-post conduct demonstrates only a lack of understanding of that transforming moment when a civilian becomes a soldier and takes on the unique rights, duties, and obligations of a member of the profession of arms.

Fortunately for policy-makers, the significance of this transforming moment was not lost on every court. Indeed, recognition of this moment's significance frequently was a factor in decisions upholding the homosexual exclusion policy. As the court noted in *Ben-Shalom*, "[c]ivilian society is not subject to those sometimes harsh and specialized military demands, and fortunately need not be, because our civilian society can depend on the military, which is."[167]

There can be no doubt that the Army, because it is an *army*, is—and absolutely must be—concerned with the conduct of its soldiers, whether off-duty or off-post. In discussing drug offenses,

Judge May, sitting on the Navy-Marine Court of Military Review, explained:

> It is not solely the immediate period prior to commencement of an infantry assault, the lifting off of a helicopter flight, the general quarters alarm aboard ship, or the dispatch of attack aircraft, which is of critical importance when gauging the effect of illegal drugs on our military forces. We must also be aware of the impact created by . . . *the decline in response to service ethics,* and the effect upon the ability of the individual marine, sailor, soldier, airman, or coast guardsman to incorporate within their conscious will the ability to respond effectively and immediately to the orders and directives of military superiors.[168]

All misconduct, whether off-duty, off-post or not, effects an incremental decline in response to service ethics. This ethical decline jeopardizes obedience in all instances as well as the discipline of the force in general.

This potential for individual and general ethical decline in the response to command inheres in all misconduct, including homosexual misconduct. Moreover, the Army must have soldiers who understand and accept their military status and its resulting responsibilities, including the need for sacrifice. Plainly, it is no defense—either to the criminal offense of homosexual conduct or to its deleterious impact on good order, discipline, and morale—to say as one former Air Force squadron commander said, "I'd always known [homosexual conduct] was against the rules . . . [but] it was my private time while I was off duty."[169] The homosexual exclusion policy clearly is not irrational on the ground it encompasses—as does the criminal law—the off-duty, off-post conduct of soldiers.

PLAINTIFFS' ARGUMENTS ON THE IRRATIONALITY OF THE HOMOSEXUAL EXCLUSION POLICY:
Conclusion

In an attempt to establish that the homosexual exclusion policy lacked a rational basis, plaintiffs argued that the policy was irrational because it punished homosexuals for their homosexuality, because it did not account for the equities surrounding individual homosexuals, and because it failed to be pro-active on the issue of homosexual rights. This last "failure" was cited in arguments that the homosexual exclusion policy was based on realities the Army should endeavor to change—that the Army should somehow attempt to counter prevailing societal attitudes toward morality, homosexual conduct, and the off-duty behavior of soldiers.

The courts finally rejected these arguments. Moreover, regardless of merit, such arguments could not carry the burden of evidence negating the Army's showing of a rational basis for the homosexual exclusion policy. Plaintiffs did not and could not show that homosexuality—as a characteristic of a group—*is* compatible with military service. Thus, every court that considered the issue found without difficulty that the homosexual exclusion policy met the rational basis standard.

CHAPTER FIVE

The Regulatory Basis for the

Homosexual Exclusion Policy

INTRODUCTION

*A critic should . . . not check a great commander's solu-
tion to a problem as if it were a sum in arithmetic. To
judge . . . it is necessary for a critic to take a more
comprehensive view.*

—CLAUSEWITZ, *On War*

The Secretary of Defense has determined that homosexuality is
incompatible with military service. This determination is not based
on the circumstances, conduct, or qualities of individual homosexuals. Rather, it is the result of applying common sense and experience to discernible characteristics of homosexuals in general and to
their conduct and then making a *judgment call*. This use of
"classwide presumptions" is recognized as both necessary and appropriate in policy-making, as in the *Gilliard* case, where the Court
held Congress was entitled to use a classwide presumption of how
custodial parents use support funds.[1]

Clearly, the determination that homosexuality is incompatible
with military service is not, strictly speaking, a determination of
fact. Rather, it is an exercise of professional military judgment in
making a broad policy *choice*. The analytical process in making

policy choices such as those underlying the homosexual exclusion policy, might be rendered in terms of five questions:

(1) What characteristics can be discerned about the circumstances and conduct of homosexuals as a class?
(2) Are those characteristics compatible with military service? That is, do those characteristics tend to make successful soldiering—for the individual and the force as a whole—*more or less likely*?
(3) What administrative burden would the Army bear if it attempted to identify individuals who were—or might be—and would continue to be exceptions to the classwide presumptions?
(4) Would the benefit of attempting to identify individual homosexuals whose circumstances and conduct are compatible with military service outweigh the administrative and other costs to the Army?
(5) What types and magnitude of risk would the Army bear for inaccurate decisions on which homosexuals are eligible for enlistment?

These policy questions can be summed up as a simple cost-benefit analysis. The Army must compare what it might gain by allowing homosexuals to serve with what it might lose.[2]

Every step in this analytical process presents complex, competing factors and interests requiring policy judgments. Yet some argue that the secretarial exclusion of homosexuals is a wrong basis for policy-making because the validity of the Secretary's judgment cannot be *proven*.

This question of whether or not policy choices must be proven—that is, with empirical data, scientific research, "hard" facts—is not new. In *Gilliard*, the Supreme Court noted that Congress had relied on a series of reasonable *assumptions, presumptions,* and *common sense propositions* in formulating a statutory entitlements scheme.[3] Based on the reasonableness of these assumptions, presumptions, and common sense propositions—though not proof proper—the

Court went on to find "the justification for the statutory classification is *obvious*."[4] Similarly, the military did not need specific proof that inoculations against disease served a "clear public interest in the military community," but an order to be inoculated was legal based on common knowledge that such shots would be highly beneficial to the health and welfare of the military community.[5]

The law clearly does not require proof of the factual merits of the Secretary's exercise of judgment in making policy choices. If the merits were considered, however, certain facts about homosexuality would pertain—even though the question of how much weight to give these facts always remains the prerogative of the policy-maker. Such facts are readily discernible from a survey of medical and other literatures documenting homosexual behaviors and homosexual culture. It is necessary to consider this evidence in the context of the rationales for the homosexual exclusion policy as set out in the Army's regulation.

THE REGULATORY RATIONALES

"The military commander has traditionally been responsible for the health, welfare, support, morale, and discipline of the members of his command, expecting in return loyalty and dedication to mission."[6] This fundamental principle of military life is reflected in the seven rationales the Army regulation sets out in support of the determination that homosexuality is incompatible with military service. These rationales are:

The presence of homosexual individuals in the military adversely affects the ability of the armed forces:

(1) to maintain discipline, good order, and morale;
(2) to foster mutual trust and confidence among soldiers;
(3) to ensure the integrity of the system of rank and command;
(4) to facilitate assignment and worldwide deployment of soldiers who frequently must live and work under close conditions affording minimal privacy;

(5) to recruit and retain soldiers;

(6) to maintain public acceptability of military service, and

(7) to prevent breaches of security.[7]

These rationales are interrelated. Each eventually points to a cumulative adverse impact on good order, discipline, and morale of the armed forces—in short, an adverse impact on combat readiness. Indeed, combat readiness should be the guiding light in any discussion of the homosexual exclusion policy, and any question that does not take combat readiness into account is a question asked in vain. Combat readiness—as demonstrated by *military success*—embraces "sacrifice of life and personal liberties, secrecy of plans and movement of personnel; security; discipline and morale; and the faith of the public in the officers and men and the cause they represent."[8]

Although not so neatly severable in reality, the regulatory rationales can be divided into four sets for purposes of discussion. Thus, the discussion will focus on homosexuality as it relates to military concerns for good order, discipline, and morale. The discussion then will turn to security concerns and privacy concerns. The final set of related rationales—"to recruit and retain soldiers" and "to maintain the public acceptability of military service"—will be discussed in Chapter Six, which considers the politics of social experimentation.

GOOD ORDER, DISCIPLINE, AND MORALE

Good order, discipline, and morale in the armed forces is a dynamic, over-arching concept. As a rationale for the homosexual exclusion policy, it is a goal in itself. As the Supreme Court itself has recognized, "no military organization can function without strict discipline and regulation."[9] For these purposes, however, good order, discipline, and morale also encompass two other regulatory rationales: "to foster mutual trust and confidence among soldiers" and "to ensure the integrity of the system of rank and command." Discussion of the potential impact of homosexuality on the good order, discipline, and morale of the armed forces will address homosexual sexual practices, health issues related to these practices, substance abuse, and related mental health issues.

Homosexual sexual practices

Throughout litigation challenging the homosexual exclusion policy, there was little dispute that homosexual acts were detrimental to the good order, discipline, and morale of the armed forces.[10] Facts about the scope, nature, and result of homosexual sexual practices are useful, therefore, in evaluating the bases upon which a policymaker could conclude that excluding homosexuals is a rational response to concerns about the detrimental effects of homosexual conduct within the military.

Research suggests there are qualitative and quantitative differences between patterns of homosexual and heterosexual activity.[11] There is ample evidence homosexuals are likely to have significantly greater numbers of sexual partners than heterosexuals. Examples in the literature include studies showing "homosexual men . . . reported a median of 1,160 lifetime sexual partners, compared with . . . 40 for male heterosexual intravenous drug users";[12] "homosexual men had significantly more sexual partners in the preceding one month, six months, and lifetime (median 2, 9, and 200 partners, respectively), than the heterosexual subjects (median 1, 1, and 14 partners)";[13] and "homosexual patients are likely . . . to have more partners . . . than heterosexual patients."[14]

It is common in the literature to find homosexuals reporting median lifetime numbers of partners in excess of 1,000. One study reported the "median number of lifetime sexual partners of the [more than] 4,000 [homosexual] respondents was 49.5. Many reported ranges of 300-400, and 272 individuals reported 'over 1,000' different lifetime partners."[15] Another study reported:

[h]eterosexual patients from all risk groups reported considerably fewer sexual partners than did homosexual men, both for the year before onset of illness and for lifetime. . . . Homosexuals had a median of 68 partners in the year before entering the study, compared to

a median of 2 for heterosexuals. . . . Homosexuals in the study had a median of 1,160 lifetime partners, compared to a median of 41 for heterosexuals in the study.[16]

In another study of 93 homosexuals, the "mean number of estimated lifetime sexual partners was 1,422 (median, 377, range, 15–7,000)."[17] It has been noted that "[l]ifetime total partners [may be] more reliably reported than either numbers of partners in the last 6 months or numbers of partners during the high period."[18]

While public health concerns may have caused some downward trend in numbers of homosexual partners, this contention has not been established conclusively. Studies continue to show the commission of unsafe sex practices. In one study, "[s]exual practices were also strongly associated with the number of sexual partners . . . For example, those practicing safe sex averaged 1.8 partners over the past 6 months, compared with 14.5 partners for those in the unsafe group."[19] Moreover, researchers have found the data suggest that "people tend to bias responses toward the socially desirably answers [note omitted]. Under the pressure of the AIDS epidemic, subjects may under-report sexual activity."[20]

Several other facts suggest there has not been a significant or long-term change in either numbers of partners or types of sexual acts practiced by homosexuals. A study published in 1990, nearly ten years after AIDS and its modes of transmission became widely known, showed "a disheartening amount of unsafe sex occurs among 18- to 25-year-old homosexual men. (This is a matter of urgent concern, as is also 'relapse' to unprotected anal intercourse among homosexuals who had switched to other practices or used condoms earlier.)"[21] Another report indicated:

[In a study of 823 homosexual or bisexual men, during] the previous 2 months, 64% had engaged in at least one sexual behavior considered unsafe. . . . This study has documented that the incidence of unsafe sexual behavior was high among gay and bisexual men in Los Angeles, even among men seeing private physicians and among men who were HIV positive or diagnosed with AIDS.[22]

A third example from medical literature reported a study where:

> Sixty percent of the total sample continued to practice receptive anal intercourse [after HIV testing]. Significant numbers of homosexual men in San Antonio, Texas, Boston, Cleveland, and Albuquerque have continued to practice unprotected receptive anal intercourse.[23]

Evidence of stable or increased patterns of incidence of sexually transmitted diseases supports the proposition that public health concerns may not have a substantial or permanent impact on homosexual sexual practices.[24] In one cohort, "[m]ost HIV antibody-positive persons in this study were identified because they had other [sexually transmitted diseases], and thus had recently engaged in sexual activity that could have resulted in the transmission of their HIV infection."[25] Again as recently as 1990, a decade into the AIDS epidemic, there were reports "[e]vidence is growing that young homosexual men are ignoring safe sex warnings . . . there had been a slight increase in the incidence of gonorrhea and syphilis, which suggested condoms were not being used."[26] Findings reported at the Sixth International Conference on AIDS also suggested homosexual sexual practices had not altered dramatically in light of public health concerns. One report from the conference noted:

> [S]ome homosexuals are relapsing into high-risk behaviors, AIDS experts said here today. . . . [H]igh-risk behavior was also a problem among younger gay men who had not yet learned the horrors of the disease first hand. The sexual practice that worries the experts most is anal intercourse, especially without a condom. It is widely regarded as the practice most likely to transmit the virus that causes AIDS.
>
> . . .
>
> Many young men appear to be ignoring advice on safer sexual practices. In another San Francisco study, Dr. George Lemp tested 816 gay and bisexual men and found that 41 percent of those aged 20 to 24 years old were infected, indicating that they engaged in high-risk sex during the mid-1980's when older gays were shunning such practices.[27]

Finally, anecdotal reports from the homosexual community suggest homosexual sexual practices, including types of sexual acts and
numbers of partners, remain an important part of homosexual culture. One report stated,

> [T]he panic of the 1980s is over. Men who once shied away from the
> bars and parties are returning for . . . "the pro-sex Nineties." . . .
> "There's an active pro-sex backlash to an anti-sexual time" . . . [Homo-
> sexuals] are again gathering for sex with strangers, the difference
> being that [at safe-sex clubs] they must use condoms. Among the 19
> percent [surveyed] who sometimes take no precautions, men under 25
> are twice as likely to lapse.[28]

One commentator observed, " '[i]t is difficult to legislate for behavioral patterns' . . . lately there had been real concern that in the past
eight years people had become tired of thinking about safe sex."[29]
Thus, relapses into risky sexual practices have been noted with
increasing frequency, as in this report:

> Dr. Ron Stall of the University of California at San Francisco reported
> that a survey of 389 gay men . . . who were trying to reduce high-risk
> sexual activities found that 19 percent reported sometimes reverting to
> risky practices. A similar survey by the San Francisco AIDS Founda-
> tion found a 16 percent relapse rate among 401 gay men who had
> vowed to practice safe sex.[30]

Another report stated a "significant proportion of homosexual men
appear to be reverting back to risky sexual behavior, threatening to
undo gains made in halting the spread of AIDS, researchers
warned."[31]

Further, reducing the number of partners may not reduce the
chance of exposure to infections, such as with hepatitis or HIV,
because of the increase over time in the size of the pool of infected
individuals.[32] Moreover, it can be reasonably presumed,

> the prevalence of HIV infection at the present time is also much
> greater in the homosexual male population than in the heterosexual
> female population. Therefore, for each random, anonymous sexual

encounter . . . , a homosexual man is more likely to contact an infected male partner than a heterosexual man is to encounter an infected female partner.[33]

Experts at the Sixth International Conference on AIDS identified several factors they said make the danger of relapse into unsafe homosexual sexual practices particularly worrisome. One of these factors: "[w]ith a large pool of infected people, estimated at up to 1 million nationwide, slippage [into unsafe sexual practices] by even a small percentage could result in a significant number of new cases."[34]

Evidence exists that homosexual acts are more likely than heterosexual acts to be committed furtively—that is, when the identity or face of the partner is hidden—and in public or semi-public, such as in rest rooms, highway rest stops, and public parks. One study published in a medical journal on sexually transmitted diseases found "[g]ay men apparently have larger numbers of different sex partners, engage in furtive sexual activities more often, and have unprotected anal intercourse more frequently than do heterosexual men."[35] Studies suggest "[m]eeting in gay bars or public places (e.g., parks, beaches, on the streets) accounted for the majority of [homosexual] encounters" for some individuals.[36] This type of sexual encounter—called "cruising"—is the subject in some travel guides. For example,

> Officials at [local] universities have been forced to consider increasing security at campus buildings that became major homosexual "cruising" spots after being listed in . . . a North American guidebook for homosexuals [that] lists buildings . . . where homosexual men can seek out semi-public illicit sex. . . . "It is not conducive [to the functions] of the university to have people engaging in homosexual activities in the bathrooms." . . . [There is concern with] the predisposition of off-campus men to frequent the building.[37]

Similar patterns of sexual liaisons while in the military have been self-reported by homosexuals.[38] One former service member told how in "the Enlisted Men's Club, the john was very active and guys would meet in there to have sex. If not right there, out somewhere, out in the fields."[39] Another former service member reported, "I laid real low for a while [in the military] and then started seeking out the 'tearoom' scene [sexual acts in public toilets]. The train stations were real good sources for that kind of stuff."[40]

A third former service member talked of his time as an officer in the Air Force and said, "I hit some [homosexual] bars and a bath or two . . . men can go out and just diddle and have a wonderful time . . . In the gay culture particularly, you do that. . . . We get physical, we get real promiscuous, and there are outlets and places where you can go and you don't even have to know your partner's name."[41] Plaintiff Dronenburg described his experiences in the military and stated,

> [T]here was a place in Seoul, it was on one of the bases . . . where people could meet, look each other over, start a conversation, or swing down the hall and *then* start into something more than conversa-tion. . . . [The USO] had facilities and you could identify other gays there, and more often than not you'd drift into the shower, and some-body would drift with you.[42]

Thus, anecdotal evidence suggests homosexuals who enter or re-main in the military by evading the homosexual exclusion policy are likely to follow the same patterns of behavior as homosexuals out-side the military.

Homosexual partners frequently are anonymous. Researchers have observed, for example, "conventional case-contact approaches [are difficult in homosexual populations] because of the large num-ber of sex partners . . . and the high frequency of 'anonymous sex,' such as in bathhouses and bars"[43] and "partner notification for patients with hepatitis B . . . has proven difficult because of the . . . large number of anonymous sex partners among many homosexual men."[44] In one study of homosexual adolescents, "in nearly one third of the cases (24/79, 30%), subjects reported 'we didn't know each other before initiating sex.'"[45] Homosexual women are less likely to fit this pattern, and research suggests they "do not often engage in sex with anonymous partners."[46]

The prevalence of partner anonymity may result from the fact "[m]any homosexuals deliberately seek anonymous sexual intercourse."[47] Case studies show patients who "had an estimated lifetime total of 4,000 sexual partners . . . visited bathhouses weekly, where he had sexual relations with an undetermined number of men . . . had had sexual relations with 100 men in bathhouses. . . . had more than 1,000 sexual partners at bathhouses and at home."[48] Epidemiological studies cite "business establishments such as bathhouses, bars, and adult book stores, [as places] where multiple, usually anonymous sexual encounters take place among a male homosexual clientele."[49]

The American Medical Association Council on Scientific Affairs has stated an individual who shows,

> a dominant pattern of frequent sexual activity with many partners who are and will remain strangers, presents evidence of shallow, narcissistic, impersonal, often compulsively driven, genital- rather than person-oriented sex and is almost always regarded as pathological. Whether such patterns of activity may yet allow appropriately satisfactory adjustments in other aspects of living is not the only issue.[50]

Researchers found data in one study that reflected "the fact that the sample [of homosexuals] has a degree of psychological disturbance, most clearly associated with higher numbers of partners."[51] Anecdotal evidence also may support a correlation between psychosociological adjustment and high numbers of partners. In *High Tech Gays*, for example, plaintiff admitted to weekly participation in promiscuous homosexual acts and to receiving treatment for an ongoing schizophrenia disorder.[52]

The anonymity and number of partners involved in homosexual acts has an adverse impact on the ability of the relative community—in this case, the military community—to control sexually transmitted diseases. Patterns of transmission of sexually transmitted disease may be affected by the size of the community and the size of the homosexual sub-community. One researcher found an association between sexually transmitted diseases among homosexuals and urban residence.[53] The study suggested:

The association between [sexually transmitted diseases] and urban residence among male homosexuals . . . is possibly a function of greater opportunity for anonymous sexual encounters in urban communities. This idea is supported by our finding that, among males, urban residents reported a larger number of sex partners per month than do those living in rural areas.[54]

Contact tracing, the primary and traditional public health response to sexually transmitted disease, is virtually impossible in these circumstances.[55] In a study on transmission of hepatitis B, "[c]ontact tracing failed, in part, because of large numbers of anonymous sexual partners among young homosexual men."[56]

Homosexual sexual practices include oral and anal sodomy, oral-anal contact, or anilingus (called "rimming"), oral-perineal contact, anal-fist contact (called "fisting"), and anal-digital contact.

Oral and anal sodomy are recognized as the most commonplace homosexual sexual practices. In one study:

Homosexuals and heterosexuals were compared for a variety of factors, including sexual practices. Homosexuals were much more likely to practice inserting penis into partner's rectum [98% of homosexuals compared with 13% of heterosexuals], inserting partner's penis in mouth [98% of homosexuals compared with 6% of heterosexuals], and inserting partner's penis into own rectum [94% of homosexuals compared with 3% of heterosexuals].[57]

Other studies found, for example, "oral sex is also a more frequent practice among gays,"[58] "oral-genital contact is a common sexual practice in [homosexual women],"[59] and the "most common sexual activity among gay men in the smaller cities was unprotected anal intercourse."[60]

Research suggests oral-anal contact, also called anilingus or "rimming," is a prevalent sexual practice among homosexuals. In one study, "92% of these men reported that they practiced receptive anal intercourse, and 63% practiced anilingus."[61] In another study the "assessed [homosexual] behaviors included unprotected anal and oral intercourse using condoms, [and] oral-anal contact."[62]

The incidence of oral-anal contact has medical significance, as demonstrated by findings that "[h]omosexual males are at high risk for acquiring hepatitis A as a consequence of promiscuity and the practice of oral-anal sex"[63] and "[r]isk factors for [hepatitis B] infection in homosexual men include . . . oral-anal sex." [64] In one epidemiological study of the transmission of hepatitis A, patient histories showed "[patient 2 was] a lesbian with a history of oral-genital and oral-anal sex with patient 1 . . . [patient 3 also] had regular oral-genital and oral-anal sex with patient 1."[65]

Anal-fist contact, called "fisting," is a high-risk homosexual practice. In one study, the relationship between high-risk behaviors and psychological status was explored by cross-tabulating "the help-seeking intention of gay men . . . with several indices of their high-risk sexual behaviors (condom use, unprotected anal intercourse, unprotected oral sex, fisting, rimming, and shared sex toys)."[66] When fisting results in medical complications, "accurate diagnosis of free perforation in rectal trauma is also important [because in] the homosexual community, fist fornication is becoming increasingly common."[67]

Secondary homosexual sexual practices include rectal insertion of objects[68] and urination on or into the body cavities of the partner.[69] The incidence of rectal insertion of objects may be demonstrated by incidence of associated rectal trauma. Research indicates "[a]norectal injuries and retained colorectal foreign bodies secondary to homosexual activities, sexual assaults, and transanal autoeroticism are common occurrences."[70] In one study, a doctor reported the "incidence of trauma to the rectum, secondary to homosexual practices, is increasing. . . . 112 patients with trauma of the rectum or with retained foreign bodies, or both, resulting from homosexual or autoerotic practices, were seen."[71] In another cohort, the "incidence of 'fisting' [and] exchange of urine . . . was low, but the incidence of other unsafe or possibly unsafe sexual behaviors was considerably higher."[72]

Research suggests "sexual practices found to be frequent among homosexual men were rare among [heterosexuals]."[73] For example, one study found "[s]exual behavior, as reflected in the number of sexual partners and history of sexually transmitted disease, was different in [the homosexual and heterosexual] groups. None of the heterosexual men participated in anal-receptive intercourse."[74]

Health consequences of homosexual sexual practices

Facts regarding the health consequences of homosexual sexual practices also are relevant to the impact homosexuality could have within the military. Homosexual sexual practices, apart from frequency of the practice or number of partners, put homosexuals at unique risk to contract pathogens and develop disease or infection.[75] Research suggests:

> [p]ossible etiologic factors [for disease in homosexuals] included multiple sexual partners, multiple sexually transmitted diseases, medication used to treat these diseases, known viruses such as cytomegalovirus and hepatitis B virus, sexual practices that involved anal contact, and use of recreational drugs.[76]

Further, "[the homosexual community] may be underserved because the homosexual is unaware of the diseases he or she is at special risk of acquiring, and therefore fails to consult a physician when seriously ill."[77]

This unique risk of acquiring disease or infection results from the fact that "fecal-oral contamination is a common occurrence in homosexual activity and is not limited to direct oral-anal contact."[78] Some case histories noted the chief mode of hepatitis A transmission "was probably fecal-oral."[79] In other cases, "[t]ransmission of . . . enteric infections probably occurs either as a result of oral-anal and oral-perineum intercourse, or when fellatio follows anal sex."[80]

Enemas or rectal douching may be used to reduce the risk of fecal-oral contamination, but these precautions do not completely protect the participants in the sexual act.[81] Moreover, enemas or rectal douching may increase the risk of viral transmission through the rectum, as seen in the medical finding that "[r]isk factors for [hepatitis B virus] infection in homosexual men include . . . rectal douching."[82]

Fecal contamination also presents problems associated with colorectal-retained foreign bodies or other rectal trauma incident to homosexual sexual practices. One report shows the complicated medical procedures that may be required, even when the patient used enemas or rectal douching before participating in the sexual act:

A controversial point in the management of [such patients] is whether or not a loop colostomy is sufficient to avoid continuing [internal fecal] contamination. . . . We have found that most of these patients had given themselves enemas before the trauma occurred. Therefore, contamination has been less extensive than expected. . . . [E]xperienced judgment may dictate the need of a double-barrel colostomy or a Hartman procedure if fecal contamination is extensive and an excessive amount of stool remains in the colon.[83]

The use of enemas may present medical complications in that "[e]ven [homosexual] men without other intestinal infections may have an altered environment in the rectum due to factors such as mechanical irritation from rectal intercourse, chemical effects of semen, lubricants, or enemas, or effects of systemic infection or immunosuppression."[84]

Research suggests "few heterosexual[s] . . . were exposed to feces during sex or had rectal trauma. None of the heterosexual men were exposed to semen."[85]

In addition to pathogenic infection resulting from exposure to feces and semen, evidence exists that homosexuals, particularly men, are at increased risk for a large number of sexually transmitted diseases.[86] In a study of 93 homosexuals, histories of sexually transmitted disease "included gonorrhea, 65.5%; hepatitis, 52.5%; amebiasis, 49.5%; venereal warts, 40.8%; phthirus pubis, 39.7%; syphilis, 36.7%; nonspecific urethritis, 26.8%; genital herpes simplex, 22.9%; shigellosis, 16.1%; giardiasis, 10.7%; nonspecific

proctitis, 10.7%; and scabies, 6.4%."[87] A study with similar findings stated:

> In addition to high rates of gonorrhea, syphilis, and hepatitis B, gay
> men have been shown to be at high risk for venereal transmission of
> anorectal venereal warts, hepatitis A, enteric pathogens, and cyto-
> megalovirus infections. The recently described acquired immune defi-
> ciency syndrome [AIDS] involving opportunistic infections such as
> Pneumocystis carinii pneumonia and Kaposi's sarcoma accentuate the
> public and personal health risk associated with sexually promiscuous
> gay males.[88]

Data exist on the differing incidence of sexually transmitted disease in the homosexual and heterosexual communities. For example, in one study, a "past history of sexually transmitted disease was
given by 160 (89 percent) of 180 homosexual men compared with 12
(46%) of the 26 heterosexual clinic patients."[89] Other researchers
noted hepatitis B infection occurred at rates of "40% to 60% among
homosexual men, compared with 4% to 18% among heterosexual
men."[90]

The problems presented by the increased risk of disease are
compounded by "[f]requent asymptomatic oral and anal infection"
and exposure to a "wide variety of agents, often difficult to diag-
nose and sometimes unidentified."[91] Syphilis, for example "may be
asymptomatic and difficult to see in the anorectal area, leaving the
person with the infection unaware of the condition."[92] Moreover,
disease and infection may have distinct or peculiar personal or
social connotations within the homosexual community. Some researchers have concluded:

> homosexual men are less preoccupied with health issues and . . .
> there is probably less guilt or venereoneurosis in homosexual clinic
> attenders. This finding does suggest that homosexual men are less
> concerned about [sexually transmitted diseases] and probably more
> likely to underplay the illness and its impact. . . . Defining oneself in
> sexual terms [as homosexuals tend to do] makes [sexually transmitted
> disease] acceptable as a risk of one's central status.[93]

Blood- and semen-borne infectious agents may be retroviruses. This means there can be a significant lapse between the time of infection and the time of seroconversion—that is, the point at which there are sufficient antibodies in the blood to mark the presence of the virus. In the time between infection and seroconversion, the infected individual can in turn infect others, even though there is no medical indication he is a viral carrier.

Blood-borne infectious agents may affect the safety of the military blood supply and increase the risk of infection through blood spatters and battlefield transfusions. One case study reported a man who contracted the HIV through a blood spatter. The facts were the man "acquired the HIV following a motor vehicle accident. . . . During the accident the patient received multiple lacerations and was covered with the blood of similarly injured and bleeding passengers."[94]

The study went on to note this "exposure through lacerated skin to the blood of persons with a high probability of being infected with HIV demonstrates an unusual mode of transmission and emphasizes the importance of HIV prevention during travel."[95] Whatever steps or measures may be taken or recommended for "HIV prevention during travel," it is unlikely the same methods could be employed to protect soldiers from each other's blood in a combat situation.

The presence of infectious agents or disease may impact adversely on the ability of the military to administer vaccinations or inoculations safely.[96] The known risk of an adverse reaction to vaccinations or inoculations has been a proper ground for discharge from the military.[97] Missed vaccinations or inoculations can impact adversely on the individual soldier, as well as on the force as a whole. In *Henderson*, a military case challenging an order to be immunized, the court found the order was legal and failure to be immunized represented a substantial threat to the health and welfare of the military community.[98] Further, "antibiotic use is frequent [in homosexuals] and may also contribute to changes in the microflora of the intestine."[99] Frequent use of antibiotics may compromise an individual's immunosuppression system and complicate medical efforts to control infection.[100]

Anal sodomy and anal-fist contact ("fisting") may result in trauma and other complications, such as prolapsed hemorrhoids, nonspecific proctitis, perirectal abscesses, penile edema, and anal fistulas and fissures.[101] One study noted, "[i]mpalement injuries from penile and fist anal fornication, resulting in mucosal hematoma, laceration, and perforation have been reported."[102] Complications, such as "[a]nal fissure, [are] seen in . . . male and female homosexuals engaging in anal sex."[103] One case study of complications arising from anal sodomy and anal-fist contact reported a fatality:

> One of these patients died. He was a 23 year old man who presented in a state of septic shock 12 hours after fist intercourse. In this patient, Fournier's gangrene developed which resulted in necrosis of the rectum and perineum, in spite of having fecal diversion performed. Multiple organ failure developed, and he did not respond to treatment.[104]

Another possible complication from anal sodomy and anal-fist contact is seen in a study noting "female to female transmission [of HIV] has only been reported in one case and suggested in another (both cases involved traumatic sex practices)."[105]

The practice of anal sodomy puts homosexuals at high risk for proctological and genital complications, including anal cancer.[106] This increased risk is statistically significant. Studies have demonstrated "82% of homosexual patients and 72% of homosexual control subjects reported [practicing anal intercourse]. . . . anal cancer risk for men who expressed a homosexual preference [was] more than 12 times that for heterosexual men."[107] A significant association between receptive anal intercourse and squamous-cell carcinoma of the anus has been noted,[108] as have a variety of other protologic and genital complications of anal sodomy.[109]

Homosexuals are shown to have an increased prevalence of intestinal spirochetosis, which may result in rectal discharge, rectal bleeding, and diarrhea.[110] This increased prevalence may be as much as 30 times the prevalence of spirochetosis in heterosexual populations. Researchers noted "[p]revious studies have demonstrated intestinal spirochetosis in rectal biopsy specimens from 2 to 7 percent of heterosexual and 36 percent of homosexual patients. . . . We observed intestinal spirochetosis in rectal biopsy specimens from 39 (30%) of 130 homosexual men but in none of the control . . . specimens."[111] Among female homosexuals, vaginitis has been cited as the preponderant medical problem.[112]

Enteric infections of, and trauma to, the rectum or anus may result in gay bowel syndrome.[113] One doctor explained that because "so many . . . conditions are due to infection of, or trauma to, the rectum or anus, . . . the term 'gay bowel syndrome' [was coined]."[114] Gay bowel syndrome is often a diagnosis by exclusion, as seen in the finding "[s]ymptomatic anorectal disease is more common among homosexual men than among heterosexuals. In approximately two-thirds of the patients, careful study will reveal an etiologic agent for proctitis. The remainder may be classified as 'gay bowel syndrome.'"[115] In general, homosexual men are "predisposed to acquiring organisms that are sexually transmitted during rectal intercourse."[116]

Medical reports show a significant occurrence of patients presenting in emergency rooms for removal of foreign objects inserted in the rectum, or for injuries incident to homosexual sexual acts.[117] The introduction to a representative report states:

A series of 101 patients with trauma of the rectum, secondary to homosexual practices, presenting at this hospital is reviewed. Two patients were injured twice. Thirty-six patients had retained foreign bodies in the rectum, 55 had lacerations of the mucosa, two had disruptions of the anal sphincter and ten had perforations of the rectosigmoid. . . . [There was one mortality].[118]

As seen, in some instances, surgery and hospitalization is required.[119] Complications may include perforations of the intestine and death from internal fecal contamination.[120]

Finally, the medical community repeatedly has recognized optimal medical care for homosexuals requires training, procedures, and often resources beyond that required to meet the medical needs of heterosexuals. The requirement for specialized medical care—

without more—is recognized as a cogent reason for exclusion from military service. For example, courts found the military's policy to exclude transsexuals was rational because, among other things, the presence of transsexuals in the service would require the Army to acquire facilities and expertise to treat endocrinological and other complications of transsexualism.[121]

The need for specialized medical care might be seen in simple cases where, without additional steps based on the patient's homosexuality, a wrong diagnosis might be reached. For example, "[i]f the examining physician is not familiar with the sexual history of the patient, pharyngeal gonorrhea . . . may be treated merely as a 'sore throat.' "[122]

The need for specialized medical care to ensure that homosexuals are adequately served—and that the Army's interest in the health and welfare of its soldiers is maintained—might be seen in increased medical requirements based on routine check-ups and regular laboratory testing for homosexuals. For example, one report advised, "it is recommended that individuals with large numbers of anonymous partners and particularly those who often practice anal intercourse obtain routine check-ups . . . [including] obtaining culture specimens from the orpharynx, urethra, and rectum.[123] The author of this report, however, recognized the "above recommendations are difficult to implement."[124]

Individuals, particularly adolescents or young adults, may struggle with acquisition of a homosexual identity. Meeting the medical and psychological needs of homosexuals requires dedication of resources and trained personnel. One researcher explained,

Although families and youths may directly consult health professionals about homosexuality, the issue more commonly surfaces as a result of other concerns such as sexually transmitted diseases, substance abuse, family conflict, academic underachievement, prostitution, and attempted suicide. Successful management [of homosexual youths] . . . underscores the importance of monitoring behavior and development and attending to frequently unmet medical needs such as hepatitis B immunization and sexually transmitted disease screening and education [for] this underserved and highly vulnerable population of adolescents.[125]

Although each military branch offers drug and alcohol counseling and rehabilitation programs, some health care providers suggest that specialized programs or facilities are required to provide adequately for homosexuals dealing with drug and alcohol abuse. For example, one report describes how a drug and alcohol program, called the Pride Institute, "aims to affirm homosexuality as an alternative life-style, to help patients understand the role their homosexuality plays in chemical dependency, and to help them learn how to deal better with the heterosexual community."[126]

If homosexuals were allowed to serve in the military, the Army would bear at least some of the financial, personnel, and morale burden of the unique risk of disease, infection, or injury that inheres in the homosexual community. Even though emergencies are infrequent compared to the incidence of disease or infection, an instance of rectal trauma, for example, shows the complicated medical issues and expensive medical procedures that might be required. One doctor noted:

> Serious injuries, secondary to homosexual acts, can and do occur, as evidenced by the mortality reported in this series. Perforations of the rectosigmoid above the peritoneal reflection can be treated by laparotomy, repair of the perforation, removal of gross [fecal] contamination by irrigation, proximal loop colostomy and appropriate antibiotic therapy. Perforations below the peritoneal reflection are challenging instances which require individualized management.[127]

Considering the serious health risks involved, the resources required to provide adequate, world-wide medical care to homosexuals in the military could be substantial. Under the military's current HIV policy, for example, a soldier who tests positive for HIV while on active duty remains on active duty with full benefits until he falls below medical retention standards. Then he may be medically retired, thus receiving a pension, life-time medical benefits for himself and eligible family members, continued participation in subsidized life insurance or other survivor benefit plans, and so on. If the cost of this compensation package averages $200,000 per soldier, the financial cost to the service is $2,000,000 for *every ten additional soldiers* who test positive for HIV.

Substance abuse

The adverse impact of substance abuse upon good order, discipline, and morale within the armed forces is self-evident. In the civilian context, the Supreme Court has stated that the regulation of drugs is so manifestly in the interest of public health and welfare it needs no discussion.[128] In the military, "even the most minor use or abuse of drugs is inimical to, and has a debilitating effect upon, the ability of the armed forces 'to perform [their] constitutional duty to fight— and to be fully prepared to fight—to protect this country's national interest.'"[129] Moreover, the military has recognized that alcohol can be deleterious to the "morale, good order, discipline, effectiveness, efficiency, and health of our armed forces."[130]

Thus, facts about the incidence of alcohol and drug abuse within the homosexual population are useful in suggesting the potential impact on good order, discipline, and morale if homosexuality was not a disqualification for military service.

Some researchers have concluded between 30 to 50 percent of the homosexual population suffers from alcoholism.[131] The estimated incidence of alcoholism among the heterosexual population is ten percent.[132]

In one study, "18% per cent of the male and 23% of the female [homosexual] respondents received scores on the Brief Michigan Alcoholism Screening Test that . . . were indicative of high risk for alcohol-related problems."[133] Homosexual women were found to have a significantly higher prevalence of alcoholism than heterosexual women,[134] and research suggests that homosexual women frequently may be concerned about their use of chemical substances.[135] In a study of 1,917 lesbians, "one-fourth reported drinking several times a week and 5% said they drink daily. Overall, the study found that 14% of lesbians worried about their chemical substance use."[136]

The homosexual bar, which functions as a place of social and sexual opportunity central to the homosexual community, may contribute to the incidence of alcoholism.[137] The community importance of the homosexual bar has been referenced frequently in the literature. The following excerpts are representative of these references:

> Alcoholism is an additional health problem for both gay men and lesbians. . . . [*It has been*] *estimated that gay men and women have a risk of alcoholism that is two to three times that of the general population. One reason for this high incidence of alcoholism among gays may be the importance of "gay bars" in the social life of the gay communities;* gay bars are the most frequently used institution for introduction into the gay community. . . . Furthermore, especially for urban males, gay bars often serve as a source of contacts for sexual liaisons. As a result, alcoholism may be associated with [sexually transmitted disease] in the [homosexual population].[138]

> Among the special problems of gay alcoholics is the gay bar, which provides social and sexual opportunities. Remedial measures are suggested that include setting up a special section at bars for serving nonalcoholic drinks and teaching gays alternate ways of finding sexual partners and socializing.[139]

Alcohol or drug abuse may contribute to high-risk behavior. Reports "suggest that drug use increases transmission of . . . sexually transmitted diseases through . . . high-risk sexual behaviors."[140] One study found "a linear relationship between the use of drugs and the kind of sexual behaviors practiced during the same time period. Clearly, men who had engaged in unsafe sex were significantly more likely to have used drugs more often than men who had engaged in safer sex."[141] Other studies attempting to identify the relationship between alcohol and drug abuse and high-risk behavior found some of "the reasons given for relapse [into unsafe homosexual sexual practices] . . . were being 'turned on,' partner pressure, the influence of drugs or alcohol, not having condoms and stress."[142]

Data similar to that reported on alcoholism are reported on the incidence of drug abuse in the homosexual community. "Estimates of serious substance abuse among male homosexual adults range as high as 30%, and . . . data suggest even higher rates among [homosexual] adolescents."[143] Medical journals provide many examples. In one study of 823 homosexual men, doctors found "[r]egarding drug use, the prevalence for most substances was high. During the prior 2 months, 6% had used intravenous drugs, 3% had shared needles, 50% had smoked marijuana, 18% had used crystal, 26% had used cocaine, and 32% had used poppers [nitrites]."[144] In another study of 93 homosexuals,

> [a]myl or butyl nitrite and marijuana were used by all patients but one. Other drugs used included cocaine, methaqualone, LSD, amphetamines, ethyl chloride, barbiturates, phencyclidine and heroin. Most patients took drugs orally or by inhalation. Only three said they had ever injected drugs either intravenously or subcutaneously.[145]

Similar findings were made with respect to other cohorts. These findings included: "[a]ll patients were homosexual or bisexual, and all engaged in anonymous sexual contact. . . . All but one . . . admitted to the use of 'recreational' street drugs;"[146] "[s]eventy-four percent of [40] patients, including all homosexual men, attended 'shooting galleries,' where anonymous multiple-partner needle sharing took place;"[147] and the "[p]revalence of [amyl or butyl] nitrite use among male homosexuals is very high . . . A guide to homosexual love-making asserts that the use of amyl nitrite has 'passed into every corner of gay life.'"[148]

Anecdotal evidence may support the reported incidence of alcohol and substance abuse within the homosexual population. Several homosexuals who had served in the military reported a problem with alcohol or drugs or a period of abuse.[149] One former Marine, a lesbian, described her time in the Marine Corps as a time when "my drinking was totally out of control. . . . I was usually hung over."[150] A former airman, a male homosexual, stated, while in the Air Force, "I began having hallucinations, probably partially from the drinking, or maybe from not drinking as much."[151] Another former service member who is homosexual stated, "Just recently, I spent a year in the alcohol abuse program . . . because I was half dead from drinking."[152]

In the case of a former Technical Sergeant challenging the homosexual exclusion policy, the plaintiff claimed that while in the Air Force he had engaged in a single homosexual encounter brought on by alcoholism and depression.[153] Similarly, at a former Army officer's trial by court-martial for sodomy, the accused raised the defense of "pathological intoxication."[154] In *High Tech Gays*, another example of anecdotal evidence from the judicial system, plaintiff claimed his security clearance was denied by reason of his homosexuality, but his clearance was denied based on evidence of past drug abuse.[155]

Emotional or psychological stress factors

Military life presents unique stress factors that may tax those who have poorly developed coping skills or who are particularly vulnerable to pressure. Not everyone has the personality or temperament to adapt successfully to military life.[156] For example, at his trial by court-martial for unauthorized absence, a sailor testified, "I was harassed by my homosexuality. . . . I could not adjust to military life. . . . I began to get a lot of mental pressures, and I started using drugs to escape some of this pressure."[157]

One researcher has observed that the "stresses of being gay in a predominantly straight society may at least in part account for the alcoholism and other substance abuse estimated to occur in 35% to 40% of homosexual persons."[158] Another researcher assessing "psychological measures" found that "men engaging in unsafe sex felt significantly less in control of their emotions."[159] In populations or individuals vulnerable to stress, the pressures of military life may exacerbate the potential for stress-related pathologies or inappropriate behaviors. Thus, facts about stress factors observed within the homosexual population are useful in evaluating the factual basis for the determination that homosexuality is incompatible with military service.

Alcohol and other substance abuse—which has a significantly higher statistical prevalence in the homosexual population—

correlates to some extent with psychiatric dysfunction, although it has not been established whether substance abuse is a cause or effect of psychiatric impairment.[160] Alcoholism has been deemed "symptomatic of more basic psychopathology."[161] One researcher states, the "association of alcoholism with other psychiatric disorders is important . . . The association between alcoholism and criminality has been reported consistently over the years."[162]

Experts note that other stress factors relevant to mental or emotional adjustment result from turmoil or anxiety in regard to sexual identity.[163] In a study of 1,917 lesbians, "[m]ore than 75% of them had sought counseling at some time, a fifth of those because of their sexual orientation."[164] Indeed, one doctor found the "very experience of acquiring a homosexual or bisexual identity at an early age places the individual at risk for dysfunction. This conclusion is strongly supported by the data."[165]

Courts, too, have been privy to the stress factors present in mental and emotional adjustment resulting from a homosexual identity. In *Baker*, for example, the court heard testimony on plaintiff's "disgust and self-loathing upon recognition of [his homosexuality], his isolation and suffering, [and] his suicidal tendencies."[166] In *Rich*, the court found "an emotional crisis stemming from confusion regarding [plaintiff's] sexual identity produced health problems."[167]

Another study reveals that in regard "to mental health, [homosexuals] appear to have some unique problems arising from the emotional toll of secrecy, disapproval, and, often, internalized shame."[168] Some data suggest "it is generally true that more persons in the . . . homosexual groups [studied] . . . regardless of medical status, experienced depressive and anxiety disorders more than did heterosexual controls."[169] A radical expression of these mental and emotional stress factors is seen in a case report where,

[t]hree alcohol-dependent homosexual men with suicidal ideation and behavior consciously attempted to contract [AIDS] as a means of committing suicide. . . . In these three patients the combination of depressed mood, social isolation, chronic alcohol use, and a discriminatory, unsupportive social climate contributed to conflicts surround-

ing sexual orientation, point the way to a perceived altruistic self-destructive escape.[170]

In some instances, homosexuality itself may be diagnosed as a personality or character disorder.[171] In *Dubbs*, the former Director of the CIA "testified that in his view the medical and psychiatric professions were in 'disarray as to whether or not homosexuality is an outward manifestation of a deeper psychological problem' . . . and that '[homosexuality] raises . . . a risk which has to be resolved in favor of the agency.' "[172] One plaintiff even argued this theory on his own behalf. In defense to a fraudulent enlistment charge, the plaintiff in *Rich* argued "the Army was required to view homosexuality as a medical defect, and the concealment of a medical defect is not cause for discharge as a fraudulent entry."[173]

Contemplation of suicide may be more prevalent among the homosexual population.[174] In a study on homosexual adolescents, the author remarked:

[other researchers on homosexual youth] concluded that their subjects experienced extreme emotional turmoil as reflected in the prevalence of psychiatric consultations (60%) and suicide attempts (by one third of the sample). . . . We previously described the stressors . . . during the acquisition of a homosexual identity and we hypothesize that these factors may predispose [homosexual youth] . . . to impaired social, emotional, and physical health.[175]

One study of 1,917 homosexual women found that contemplation of suicide, frequently acted out in suicidal attempts, occurred in more than 50% of those in the study group.[176] Twenty percent of the 1,917 lesbians in the study had attempted suicide at least once.[177]

The incidence of suicide contemplation among homosexuals may be supported by anecdotal evidence. Several homosexuals who evaded the homosexual exclusion policy and entered the military have reported either an episode of suicide contemplation or a suicide attempt, generally related to their homosexuality.[178] In *Baker*, plaintiff testified he had suffered suicidal tendencies as a result of recognizing his homosexuality.[179] In United States v. Varraso, a trial by court-martial on charges of murder, Varraso's statement was entered into evidence:

> [The victim] and I discussed committing suicide together, due to my depressed state, because of consent conflicts with my lesbian lover and company personnel. . . . [She was under the influence of] quaaludes, tylenol #3 [codeine], and lots of beer. . . . All I drank that day was six beers, between 9:00 A.M. and 3:00 P.M.[180]

Attempts to maintain social norms that impose profound lifestyle changes on homosexuals—such as norms imposed by the military community—could be an additional stress factor that might result in inappropriate coping behavior.[181] Researchers indicate this clash of social norms—and the imposition of norms from outside the homosexual community—can be quite stressful. One example of a clash of social norms is the tension between the homosexual community's emphasis on sexual freedom and the public health requirement to refrain from unsafe sex. It has been observed:

> Transmission of AIDS through sexual contact or body fluid often leads to self-imposed or physician-recommended limitation of social and intimate contacts. Guidelines for safe sex among homosexuals may require marked alteration of sexual behavior and the need to reveal medical status to partners. *Pressure to accept a monogamous relationship or celibacy may be difficult.*[182]

In the Army, this clash of norms would be exacerbated because of the military's policy requiring HIV-positive soldiers to inform partners of their status and engage only in "safe sex." Soldiers have been court-martialed and convicted for violating the military's "safe sex order," including the order to inform partners of their HIV status.[183] Nevertheless, a report in the *Journal of the American Medical Association* indicates "a sizeable minority of people engaging in sexual activities that may transmit HIV did not intend to tell their partners if they are HIV positive."[184] The report added that the "lack of intention to communicate positive test results to nonprimary partners may be attributable to not knowing these individuals' names."[185]

Anecdotal reports demonstrate the stress factors that may occur for homosexuals within a military environment. Many individuals who served in the military despite the homosexual exclusion policy noted the unique stress their homosexuality caused while they were in the military. One former service member said, "I can't emphasize enough that we [homosexuals] lived in absolute terror of being found out. It would have been a terribly overwhelming experience. . . . It just isn't for everyone, the military. There is too much one must hide in order to stay in undiscovered. It's not worth it."[186] Another states, "In order to serve in the U.S. military, lesbians and gay men must live a lie and be in constant fear of exposure. That is destructive to the gay spirit."[187] A third former service member gave this advice: "I say [to other homosexuals], 'No! Keep the hell away from the military. . . . ' The risks are too great. If you're going to live and have a relaxed, happy, normal existence—[joining the military is] the worst thing to do."[188]

Simply allowing homosexuals openly to identify themselves as homosexuals within the military would not reduce this type of stress. In fact, it might increase stress by fostering suspicion of homosexual acts.[189] So long as homosexual acts and conduct prejudicial to good order and discipline are proscribed, the potential for stress and fear of discovery would remain for some homosexuals. Indeed, the Dutch military has been open to homosexuals since 1974. Nevertheless, one Dutch sailor "complained that trying to keep secret his homosexuality is hurting his work performance. 'I drink too much because sometimes I think I am going crazy, but life is rotten. It is only when I go home at the weekend that I can be more or less myself,' the sailor said."[190]

In addition to endemic mental or emotional stress factors in the homosexual population, physical health status, such as repeated incidents of infection or disease, can impact adversely on mental health status. One researcher found:

[Patients] with higher numbers of previous infections [with sexually transmitted diseases] . . . tended to deny life stressors more, to attrib-

ute their problems to the illness, and to display significantly higher
levels of disease conviction (including symptom preoccupation), of
affective disturbance (anxiety and sadness), and of exaggeration of
symptoms.[191]

The requirement to divert time and energy to physical or mental
health problems may be an additional stress factor for the individual
as well as for the relevant community, in this instance the military. In
Doe v. Alexander, for example, a case discussing the exclusion of
transsexuals from military service, the court stated,

> The regulation here at issue is . . . designed to assure the enlistment of
> personnel free of medical conditions which would cause excessive lost
> time from duty, medically adaptable to global geographic areas and
> capable of performing duties without aggravation of existing physical
> or medical conditions.[192]

This regulatory design is critical because, as a practical matter in the
Army, there are no "vacancies"—the mission simply must be accomplished. If one soldier is at sick call, another soldier has to pull
the sick soldier's duties in addition to his own. In chronic cases, this
has implications for the mission as well as morale.

Impact of homosexuality on good order, discipline, and morale: conclusions

Review of medical and other literature documenting homosexual
sexual practices, health consequences of homosexual sexual practices, substance abuse, and observed emotional or psychological
stress factors is useful in evaluating whether or not there is a factual
basis from which the Secretary rationally could conclude homosexuality is incompatible with military service.

   Despite the theoretical distinction between orientation and behavior, homosexual activity is not a remote possibility within the homosexual population. The health consequences of homosexual activity
may adversely affect the homosexual soldier. Homosexual activity
may increase the rate and scope of disease spread within the mili-

tary community. It may compromise the military blood supply and burden military medical resources as well as other soldiers, who must perform the duties of those who are sick or absent because they are seeking medical care. The potential effect of military stress factors on emotional and psychological stress factors observed within the homosexual population also poses serious problems for both the individual and the Army.

Clearly, the observed scope, nature, and result of homosexual sexual practices, as well as the observed stress factors within the homosexual population, support the judgment that homosexuality could have an adverse impact on good order, discipline, and morale within the armed forces. Similarly to the Supreme Court's finding in Dallas v. Stanglin (a case challenging a statute that excluded adults from dance halls catering to teenagers), it is fair to say "[the Army] could properly conclude that [excluding homosexuals from military service] would make less likely illicit or undesirable [soldier] involvement with alcohol, illegal drugs, and promiscuous sex."[193]

MILITARY SECURITY CONCERNS

Some perceive the security concerns of the homosexual exclusion policy as a one-dimensional "objection that because of their alleged susceptibility to blackmail [homosexuals] present potential security risks."[194] Thus, the simple reply is that this objection "could be forever eliminated if the military were to lift its ban on homosexuals—thereby immunizing [them] against blackmail."[195]

Security concerns are not so one-dimensional, however. Security threats are uniquely informed and impelled by the complexities of human nature and the human psyche—they play on the fact that every person is "just human." Thus, answers to broad-ranging threats to security are not as simple as merely changing military accession or retention standards in an attempt to "immunize" homosexuals from hostile intelligence activity. Moreover—as is usual in a *policy perspective*—the homosexual exclusion policy is not based on the individual homosexual as a security risk. The inquiry goes to whether or not *homosexuality* presents, or increases, security risks to the force overall.

Security concerns are much broader than the passing of secret documents or other sensational compromises of classified information. The military discipline of "operational security," for example, recognizes that every soldier, regardless of duties or clearance, has information that may be useful to hostile agents.

While there are myriad scenarios, one example of rudimentary, yet perhaps pivotal information a soldier may possess is medical information—perhaps what shots he received before he was deployed. During Operation Desert Storm, for example, *Army* reported, "[b]ecause of operational security concerns, [the Department of Defense] did not confirm the specific diseases against which troops are to be inoculated . . . [because] once Iraq is aware of the measures to be taken against certain diseases, [biological warfare] strains could be altered or added to counter prevention."[196]

Operational security long has been recognized as a principle of military engagement, thus as a critical component of combat readiness.[197] Further, the Army reasonably can conclude that anything which increases hostile intelligence activity is detrimental to the force, without regard to whether or not that activity results in actual breaches of security.[198] This perspective was evident in *High Tech Gays* where the court noted that the Department of Defense established that homosexuals are targeted by counterintelligence agencies and, because homosexuals are targeted, the Department subjects homosexuals to an expanded investigation.[199]

Thus, the DoD policy is based on the *fact of the targeting*, not on any presumption about homosexuals' susceptibility to the goal of that targeting. This rationale also was seen in *Krc*, where the agency's Director of Security testified he did "not want to be responsible . . . for placing [a homosexual employee] in an environment where he could be *subjected to* a hostile intelligence approach."[200]

Hostile intelligence activity presents multiple layers of threat to security. The potential for blackmail, or other forms of coercion, based on sexual information is recognized.[201] In *High Tech Gays*, plaintiff, an employee of a defense contractor and an *open* homosexual, admitted he had been the subject of blackmail attempts because of his homosexuality.[202] In an interview with a former Army officer, he stated:

[I]n my last duty assignment [as a Military Intelligence officer], it wasn't an issue of my homosexuality so much that made me leave the Army but an issue of compromise. I was being blackmailed into doing things because of it. And that's when I refused to be blackmailed. I said, "Fine, I quit" [active duty and joined the Army Reserve].203

A relationship between blackmail and homosexuality is acknowledged in other national cultures as well. In Britain, for example, a commentator opined, "[t]he law's disapproval of homosexuality is by and large supported by public opinion. . . . it is only realistic to accept that an active homosexual must run a certain risk of blackmail, all the more so if he occupies a sensitive position in society [such as a judge]." 204 In France, a man "claimed that his superiors had asked him to recruit young prostitutes of both sexes in an attempt to discredit leading public figures with suspected homosexual or pedophiliac tendencies."205

Indeed, blackmail based on homosexuality is not solely the province of hostile agents, nor solely the concern of soldiers. Far from being an "old chestnut," forms of blackmail based on homosexuality are on the rise. One need only to look to the phenomenon of "outing." Outing—the public identification of homosexuals, usually by other homosexuals, against their will—frequently has been likened to blackmail. One commentator states:

Last year, openly gay Democratic Rep. Barney Frank of Massachusetts threatened to name homosexual Republicans in Congress in reprisal for what he said was an innuendo campaign to insinuate that Rep. Thomas S. Foley of Washington, then seeking the House speakership, was gay. The threat sounded like blackmail, but it was perfectly justifiable blackmail in Mr. Frank's eyes. He said that some anti-gay Republicans had "forfeited their right to privacy."206

A spokesman for the homosexual community explained, "[o]uting has engendered a deep rift in the gay and lesbian community . . . The homosexual rights establishment . . . stands firmly by the principle that . . . '[i]nvoluntary disclosure [of one's sexual orientation] is a coercive action.' "207 The military is not the only community concerned about the threat of involuntary disclosure of an individual's sexual orientation or sexual conduct. This concern is common, even in the "open" environment in the entertainment business:

> Not surprisingly, [outing is] causing anger, anxiety and fear in Hollywood, where a sexy, heterosexual image seems crucial to many lucrative careers. "If the public knows you're gay, it has a whole different perception of you," says one actor, who is gay and worried.[208]

Outing may fade as a technique for homosexual militancy. Nevertheless, it demonstrates an enduring principle, one policy-makers are entitled to consider: many people—regardless of sexual orientation and for a variety of reasons—do not want their private sexual lives discussed publicly. A stunning example of this is seen in the Dutch military, where homosexuals have served openly since 1974. Even though a government-wide edict prohibits discrimination on the basis of sexual preference, a researcher studying the Dutch military found "[m]any homosexual sailors . . . take steps to hide their sexual orientation. . . . One sailor [said]: 'Homosexuality is not accepted at all [in the Royal Dutch Navy]. It is associated with everything that a person in the navy should loathe.' "[209]

Blackmail, however, is a worst case scenario, one that usually arises only if and when a relationship—between the hostile agent and the targeted individual—turns adversarial. Thus, while blackmail is an important security concern, it is only one factor in the complex of threats presented by hostile intelligence activity.

Compromising relationships between hostile intelligence agents and soldiers begin with identification of individuals who are perceived as vulnerable. These individuals then are targeted with positive appeals, including enticement "with money, sexual favors, [and, for example,] treatment for alcoholism."[210] Targeting involves the "exploitation of what [hostile] agencies consider to be human weaknesses, indiscretions and vices."[211] Targeting "might include surveillance of [vulnerable individuals] and their associates."[212]

The technique of reaching individuals through their partners is recognized and common. In *Dubbs*, the court found hostile intelligence agents attempted "to place [individuals] in circumstances in which their conduct could lead to arrest or other sanctions or otherwise influence their actions through direct or indirect pressure on *them or their partners.*"[213]

Civil and criminal investigations regarding homosexuality also are influenced by the subject's and witness's loyalty to partners. In Ashton v. Civiletti, for example, a mail clerk with the Federal Bureau of Investigation was named by a casual homosexual partner who was under investigation by the Navy for homosexual acts.[214] In *Dronenburg*, plaintiff initially denied homosexuality; the Navy then produced sworn statements of plaintiff's partner.[215] In a different report of investigation plaintiff stated, "I do not want to identify any persons with whom I have been involved in homosexual activity or further describe the type of homosexual activity."[216]

Besides the threat of involuntary disclosure of one's homosexual preference or acts and the direct or indirect pressure exerted by partners, the doctrine of hostile intelligence agencies, such as the former Soviet Intelligence Service (the KGB), instructs agents to attempt to

> identify those who are ideologically sympathetic, experiencing career difficulties, unsuccessful in social relationships, experiencing problems with narcotics, alcohol, homosexuality, or marital difficulties. . . . [N]o one trait may be sufficient, and . . . the KGB is encouraged when these traits are found in combination. . . . *The KGB is not primarily interested in homosexuals because of their presumed susceptibility to blackmail. In its judgment, homosexuality often is accompanied by personality disorders that make the victim potentially unstable and vulnerable to adroit manipulation.*[217]

Thus, some hostile intelligence agencies view homosexuality as a marker for other exploitable traits. Evidence exists that "[c]ertain hostile intelligence services regard homosexual behavior as a vulnerability which can be used to their advantage."[218] Indeed, in *Krc*, the Director of Security testified, "[plaintiff's] homosexuality would make him extremely vulnerable to hostile intelligence approaches."[219]

Further, the homosexual may feel disenfranchised or alienated, especially under the stress of military life and the potential sanctions for homosexual conduct.[220] The district court in *Baker* found, after expert testimony,

> the existence of these criminal laws [proscribing sodomy], even if they are not enforced . . . , does result in stigma, emotional stress and other adverse effects. The anxieties caused to homosexuals—fear of arrest, loss of jobs, discovery, etc.—can cause severe mental health problems. Homosexuals, as criminals, are often alienated from society and institutions, particularly law enforcement officials.[221]

The court in *High Tech Gays* found that homosexuals interest KGB handlers "since the homosexual frequently is shunted by society and made to feel like a social outcast. Such a personality may seek to retaliate against a society that has placed him in this unenviable position."[222]

The concomitant resentment or bitterness the homosexual might experience may be projected back at the Army or at society as a whole. Plaintiff Hatheway, for example, was a lieutenant in the Army's Special Forces. He was court-martialed, among other charges, for committing sodomy with an enlisted man in the barracks. Hatheway described an incident after his trial:

> I got a little crazy at the end, after the trial itself. I slammed open the door of the adjutant's office and . . . I was screaming at him; then I turned, went into the colonel's office, and repeated my act. . . . I berated him . . . I think he was very taken aback by the bitterness with which I reacted to him. All he could say was, *"Whatever you do, don't be so angry as to tell the whole world of what you know Special Forces does to protect the free world."* What he had done with that statement was to . . . *put his burden of national defense on me, so I told him to f--- off.* I just flipped him the bird and walked out.[223]

Whatever the cause of the potential resentment or bitterness, the Army must be concerned with—and take steps based on—the fact that it exists.

Moreover, disclosure of one's homosexuality does not "immunize" a soldier from hostile intelligence targeting. The argument that homosexuals would no longer be security risks once the ban were lifted assumes first that most homosexuals *desire* to disclose their sexual preference. In one case, plaintiff feared disclosure of his homosexuality; he stated he would rather terminate his employment with the Air Force than have his homosexuality exposed during a security clearance background check.[224] One study has noted "there are selected groups who are especially susceptible to . . . anxiety. These groups include persons . . . who fear . . . the disclosure of belonging to a group [at risk for HIV infection]."[225] In a particularly tragic case, a soldier who donated HIV-infected blood stated he "was aware that, as a practicing homosexual, he was in a high risk group for AIDS but gave blood anyway because he did not want anybody to think he was 'gay.'"[226]

Even if a soldier revealed his homosexuality, general awareness of his sexual preference might *facilitate* targeting. In *High Tech Gays*, for example, the court discussed the evidence presented by the Department of Defense (DoD):

> The DoD . . . presented . . . sworn statements of Sergeant Lonetree . . . relating one meeting with his Soviet control Sasha (later identified as . . . an officer of the KGB), where Sasha specifically inquired as to homosexuals. [Lonetree stated:] "During a meeting with Sasha, he asked me to tell him who were the homosexuals, drunks and people who were exploitable. . . . *He also asked for other Marines . . . who might be queers, dopers or drunks* . . . [on another occasion] Sasha also wanted to know *which Marines had problems with alcohol or drugs or those who were homosexual.*[227]

In *High Tech Gays* itself, one plaintiff, who was employed by a defense contractor and who was openly homosexual, admitted he had been targeted because of his homosexuality.[228]

Open homosexuals also remain vulnerable to coercion based on loyalty to (or pressure by) partners who do not want their own homosexuality exposed.[229] As one court explained, it was "not

irrational for the FBI to conclude that the criminalization of homosexual conduct coupled with the general public opprobrium toward
homosexuals exposes many homosexuals, even 'open' homosexuals, to the risk of possible blackmail to protect their partners, if not
themselves."[230]

Moreover, disclosure does not preclude the possibility that the
presence of homosexuals in a military unit will cause increased
hostile intelligence activity, which is a harm to be avoided in its own
right. It is well settled in military law that:

> The commander of a military installation is responsible for the mainte-
> nance of good order and discipline aboard that installation. The pro-
> scription against solicitation is concerned not only with the *prevention
> of the harm* that would result should the inducements prove successful,
> but with protecting those aboard military installations from being
> *exposed to inducements* to commit or join in the commission of
> crimes.[231]

As one military court put it, "criminal sanctions may be adequate in
civilian circles, but preventive measures must be taken in the military."[232]

Likewise, military security interests are best furthered—so far as
is possible—by prevention of the potential harm caused by increased
hostile intelligence activity and protection of the force from exposure
to inducements to breach security. When protecting "those aboard
military installations from being exposed to inducements to commit
or join in the commission of crimes," it is immaterial whether or not
homosexuals are more likely than heterosexuals to compromise security. The potential danger to security lies in the fact the *total effect* of
increased targeting logically might be increased security breaches,
and most certainly would be decreased morale in the command.

One final point bears on security concerns supporting the homosexual exclusion policy. The standard for granting a security clearance is not strictly applicable to this discussion of how security
concerns are implicated by the homosexual exclusion policy. However, the standard is illuminating.

For one thing, to be commissioned as an Army officer one must be eligible for security clearance to at least the "SECRET" level. Thus, if homosexuality is a factor in security clearance determinations—as it is[233]—then homosexuality also is a relevant factor in excluding homosexuals, at least from the officers' corps. Moreover, the legal relationship between officer commissions and security clearances raises a separation of powers issue, because military commissions are granted by the President, who has "unreviewable discretion over security decisions made pursuant to his powers as chief executive and Commander-in-Chief."[234] Thus, even if the homosexual exclusion policy were revoked, homosexuality still would be a potential basis for exclusion under existing procedures for granting security clearances.

A second way in which the security clearance procedure illumines the rationale supporting the homosexual exclusion policy is the standard itself. In general, a security clearance "may be granted or retained only if *'clearly consistent* with the interests of the national security.' "[235] For example, the court in Doe v. Weinberger noted a homosexual's indiscriminate pattern of sexual activity led to an agency decision that his access to classified information was *"not clearly consistent* with the national security."[236] In the context of the Central Intelligence Agency, the "clearly consistent" standard has been seen as congressional recognition of "the need for the Director to have broad power to terminate CIA personnel for *even the slightest security risk.*"[237]

This high standard illustrates the gravity of the risks involved when the security of the nation is open to compromise. The security concerns addressed by the homosexual exclusion policy also involve serious risks. These documented risks clearly are another factual matter in the calculus behind the homosexual exclusion policy.

PRIVACY CONCERNS

Another basis for the homosexual exclusion policy is the exclusion of homosexuals "facilitate[s] assignment and worldwide deployment of soldiers who frequently must live and work under close conditions affording minimal privacy."[238] This rationale addresses the privacy concerns of other soldiers that would be implicated if homosexuals were allowed to serve in the military. One manifestation of cultural privacy expectations is gender segregation. The fact is, in most non-private settings, gender segregation is the norm in sleeping arrangements, shower facilities, and bathrooms, as well as in some occupations, such as clothing sales.

Gender segregation is based on two presumptions. The first presumption is that people have a *sexual* preference for persons of the opposite sex. The second presumption is that people should be allowed to choose to whom—by exposing their bodies or by engaging in intimate bodily functions—they expose an aspect of their *sexuality*. Since sexuality is not in issue when heterosexuals are segregated by gender, the privacy implications of bodily exposure and intimate bodily functions are not in jeopardy.

Society considers the privacy entitlements protected by gender segregation so important that jeopardizing them can be a crime. In United States v. Johnson, for example, the accused secreted himself in the women's latrine in order to watch women use the facilities.[239] The court found the accused's "act constitutes an invasion of privacy of those observed" and "evinces a wanton disregard for a normal standard generally and properly accepted by society, in addition to bringing discredit upon the Armed Forces in the eyes of the victim and her sister."[240]

Courts have recognized the privacy considerations protected by gender segregation in several settings. In urinalysis cases, such as National Treasury Employees Union v. Von Raab,[241] courts note as a given in the urinalysis procedure that the observer should be of the same sex as the person providing the urine sample.[242] In the instance of body searches, even visual only, one factor in the reasonableness of the search is whether it was performed by an officer of the same sex as the individual who was searched.[243]

In prisons—where personal freedom is at the minimum—the law still recognizes inmates of both genders have some privacy entitlements to same-sex guards in certain situations, such as when using the shower or bathroom facilities.[244] Unlike the infrequent situation of undergoing urinalysis, and unlike the distant relationship between guards and inmates, the privacy considerations protected by gender segregation would be at issue even more among people actually living together in close conditions. Clearly, society finds appropriate and chooses to preserve the privacy considerations historically protected by gender segregation.

Similarly to society at large, the Army often depends on gender segregation to protect a person's privacy entitlements while at the same time allowing for efficient, cost-effective processing and housing of large numbers of soldiers. Gender segregation, in many situations, also has clear utility for preserving good order, discipline, and morale.

The protection and restraint provided by gender segregation breaks down, however, when the underlying presumption that individuals sexually prefer the opposite sex is invalid, as in the case of homosexuals. In situations of gender segregation, soldiers rely on the fact of segregation and proceed accordingly: they may undress more freely, remain only partially dressed, sleep in fewer clothes, and be less on guard for potential sexual assaults. One former male sailor, for example, described his "techniques" while on board ship in the Navy:

> one of the techniques, which I outrageously developed, was merely crawling in with [other men] and engaging them in sex and leaving them as if nothing had ever happened. You don't say anything; you pretend it never happened. And so long as you never discuss it, it never happened.[245]

Not every homosexual, of course, seeks to be as "outrageous" as this former sailor. Still, sexual assault is one of the harms society intends to address through gender segregation.

For the homosexual man, for example, the Army's preinduction physical may be viewed as "pretty spectacular, I mean, 'cause you've got three hundred naked men in one room."[246] But the Army does not intend preinduction physicals, open bay barracks, group shower facilities, or mass inoculations to be "spectacular," only efficient. Nor do soldiers of either gender deserve to be stripped unwittingly of their right to choose to whom they reveal themselves in a sexual context. Once this happens the harm is done. As a matter of law, the privacy violation does not depend on any acting out of sexual attraction toward others. It is complete the moment privacy is breached.[247]

Thus, the only way to maintain the protection and discipline—and, as will be seen, public acceptance of and respect for military service—that is otherwise achieved through gender segregation is either to exclude homosexuals or to accommodate them, if such is possible.[248] The Army need not attempt accommodation. It "could choose to accommodate [homosexuals] or to accommodate privacy interests of [soldiers] and security interests of the institution."[249] For purposes of evaluating the rational basis for the homosexual exclusion policy, however, the very fact that accommodation efforts would even be required supports the judgment that homosexuality is incompatible with military service.

SUMMARY:
Practical Perspectives on the Homosexual Exclusion Policy and the Rational Basis Test

The homosexual exclusion policy, like any other solution reached by a military commander, cannot be "checked as if it were a sum in arithmetic." Rather, one must take a comprehensive view of the factors that might go into the commander's policy equation. There is ample available evidence from which the military reasonably could conclude that allowing homosexuality within the armed forces would present substantial risks to good order, discipline, morale, security, and to the privacy entitlements of other soldiers. One rationale may be stronger than others. The importance—or relative weight—of the separate rationales may ebb and flow over time, or vary from station to station, or change from assignment to assignment or war to war. Nevertheless, as the basis for a broad policy choice, the convergence of these several critical military concerns—good order, discipline, morale, security, and privacy concerns—certainly supports the secretarial determination that homosexuality is incompatible with military service. The Army need not try to fine tune policies to minimize risk to combat readiness. Rather, it can elect a solution that avoids risk altogether. This is all the homosexual exclusion policy represents: the policy judgment that it is better to exclude homosexuals than to risk the potential detrimental effects of homosexual conduct within the military.

CHAPTER SIX

The Politics of Social Experimentation

INTRODUCTION

*I don't know what effect they will have on the enemy, but,
by God, they scare me.*

—THE DUKE,
upon review of his troops

Throughout history the call to arms has resulted in the assemblage of armies great and small, sharply skilled and successful, or inept and therefore hopelessly doomed. As it enters the last decade of the century, the American military is the most highly trained, highly disciplined, highly motivated, and best equipped armed force in the history of the nation. This very successful program, however, has led some to believe that this Army—in addition to being a warfighting force—should also be the engine of social change, the "leader in social experimentation and . . . adaptability to changing community standards,"[1] the revisionist of the "wisdom and the values of a society that has allowed [the military] to be so wrong [about homosexuals] for so long."[2]

This opinion (that the Army has a *social* as well as a *military* mission) came into prominence at the same time the military was

132

becoming more and more "civilianized" in the public conscious-
ness. Despite recruiting slogans proclaiming, *"It's not just a job!,"* the
absence of war and the advent of the all-volunteer force made it
increasingly easy to lose sight of what the Army is all about. In
some circles, the Army was viewed as a sort of benevolent welfare
agency. In others, it was a college for professional or career development. Over time, some social theorists frankly presumed that the
Army existed simply to provide a job to everyone who wanted one,
regardless of the requirements of "soldiering." In point of fact,
however, the Army exists for only one reason: to preserve the peace
and security, and to provide for the defense of the United States in
its unique position of leadership in the international community.[3]
Thus, the end result of personnel decisions must be, not some social
utopia, but no more and no less than an *army*—and that army must
be the best in the world.

The military strength of the United States, moreover, should not
be taken for granted, nor should the complexity of the task of
fielding a successful military force be underestimated. As the Chief
of Staff of the Army has noted, the Army entering the 1990's—

> is the finest fighting force this nation has ever fielded and the best in
> the world today . . . [But] [t]his Army did not come about by accident.
> It is the product of a comprehensive and visionary plan . . . principles
> that are the benchmark by which we measure every proposal and
> every program. . . . These imperatives include an effective warfight-
> ing doctrine . . . and an unbending commitment to a quality force.[4]

Advocates of an army of social change, however, maintain that
accomplishing social goals is as important for the Army as accomplishing military goals. These advocates have advanced several proposals to modify the homosexual exclusion policy. Such
models attempt to address the problems of homosexuality within
the military without excluding homosexuals. The proposals include a "disclosure model," a "decriminalization model," and an
"accommodation model." This chapter will explore the operation
and logical soundness of each.

THE "DISCLOSURE" MODEL

The disclosure model relies on the notion that the problems of homosexuality in the military can be addressed by allowing homosexuals to serve openly, while the Army relies on the criminal justice system to prevent homosexual conduct. In this model, the homosexual is required to declare his homosexuality openly.[5] Then he is required to acknowledge affirmatively that homosexual acts are criminal offenses under the Uniform Code of Military Justice.[6] Before discussing the efficacy of this model, a preliminary observation must be made about the dynamic it creates for the soldier and the Army.

The procedure described in the disclosure model—open declaration of homosexuality followed by affirmative acknowledgment of criminal proscriptions of sodomy—has a serious disadvantage. It puts the homosexual soldier in an openly adversarial role with the Army from the moment he puts on his uniform. Indeed, disclosure of homosexuality may place the soldier in a adversarial role with the larger homosexual community, which prefers anonymity. Plaintiff Ben-Shalom described this relationship within the larger homosexual community after she challenged the military's homosexual exclusion policy. She stated:

> [T]he lack of support from the gay community doesn't help the pain [of being excluded from the military]. There was no support whatsoever. I've taken more harassment and more hurtful things from the gay community than I ever have from heterosexuals. . . . People didn't want to be seen with me in public. I don't understand why. . . . But if you're seen with a queer, walking down the street, well, my God, you must be a queer too. . . . It's been very solitary.[7]

The disclosure model says to the soldier that the Army has allowed him into the service but nevertheless lacks confidence in his ability to serve successfully. This procedure would, in essence, officially stamp homosexual soldiers as probationers. This official probationary status—almost like a notice to the soldier that says,

"We are watching you"—clearly is not the best foundation for loyalty—which must be from the Army to the soldier as much as from the soldier to the Army. Such an arrangement would severely compromise the *esprit* that is essential to good order, discipline, and morale.

Moreover, this result—that disclosure could make matters worse rather better for the Army—highlights the fact that disclosure of one's homosexuality serves no useful *policy function* apart from the homosexual exclusion policy. Disclosure has no impact on the validity of the regulatory rationales for the policy. Disclosure would not negate—and might exacerbate—military security and privacy concerns. Disclosure would not affect one's sexual conduct, use of drugs or alcohol, or coping skills. In fact, as seen in the experience of Ben-Shalom, described above, disclosure may well have a negative impact on behavior or social adjustment.

However, whether or not the model might be effective in addressing the problems of homosexuality in the military depends on two factors. These factors are the willingness of homosexuals to disclose their homosexuality and the practical effect of criminal sanctions on homosexual conduct.

Disclosure of one's sexual identity is an intensely personal decision. It is not at all clear that lifting the "sanction" for homosexuality would result in one-hundred-percent disclosure, or even in disclosure by the majority of homosexuals who presented themselves for enlistment. In *Dubbs,* for example, plaintiff did not disclose homosexuality even though it was not an automatic disqualification for employment with the CIA.[8]

A reluctance to reveal facts about one's sexual life is natural. People probably would be just as reluctant to answer questions about adultery, promiscuity, or even frequency of sexual intercourse with their spouses as they would to answer questions about their homosexuality. Even in dire circumstances, people often are traumatized further by the public airing of their private sexual lifestyles. For example, the "diagnosis [of AIDS] may force the patient's identification as a likely member of a stigmatized minority . . . Families may abruptly learn of a lifestyle they find difficult to accept. In the largest risk group, homosexual and bisexual men, the diagnosis may create *a crisis in which an otherwise private sexual preference is revealed."*[9]

The practical effect of criminal sanctions on homosexual conduct is also problematical. The practicability of criminal sanctions depends on effective deterrence as well as *cost-effectiveness.*[10] In *Harper*, prison officials restricted access to materials from the North American Man-Boy-Love Association (NAMBLA).[11] The court noted testimony that "searches for NAMBLA materials which have been introduced to the prison are not as effective as forbidding their introduction in the first place. This factor thus favors upholding the regulation."[12]

Harper, then, recognizes the common sense principle that avoiding a problem is more advantageous than attempting to solve the problem once it exists. The disclosure model is deficient to the extent that the cost of taking a "wait and see" attitude toward the homosexual's potential to commit homosexual acts—the cost in terms of combat readiness as well as financial and other resources—outweighs the benefit to the Army of permitting homosexuals to serve.

In the disclosure model, the homosexual soldier is required to sign a statement acknowledging that sodomy is an offense under the Uniform Code. The rationale: The Army is assured that homosexual soldiers "know the rules."[13] Essentially, this statement of acknowledgement is the equivalent of a pledge of celibacy on the part of the homosexual soldier. Celibacy, however, is widely regarded as an unrealistic standard of behavior, particularly among homosexuals. Further, it has not been observed as a statistically significant feature of either homosexual or heterosexual communities.[14]

Indeed, the concept of monogamy—much less celibacy—has been rejected by some homosexuals as an expression of "pervasive heterosexism" and, thus, as a "marginalization" of one's homosexual identity.[15] Researchers have observed that "[g]uidelines for safe sex among homosexuals may require marked alteration of sexual behavior. . . . Pressure to accept a monogamous relationship or celibacy may be difficult."[16] Attitudes toward monogamy and celibacy are evidenced in the homosexual community's reaction to a book titled "Faggots," written by Larry Kramer, a homosexual and well-known homosexual rights activist.[17] The book has been described as:

a scathing critique of sexual promiscuity among homosexuals that sold more than 440,000 copies. It also made Kramer public enemy No. 1 in some gay circles. The author said his book had only suggested that homosexual men look for love, instead of random sex. But *critics felt he was denigrating the gay liberation movement.* When the AIDS crisis erupted in 1981, the same tension emerged. *Kramer called for safe sex policies, while a multitude of homosexual leaders angrily resisted, saying such recommendations were premature and punitive.*[18]

In short, it is reasonable to conclude, as the court in Baker v. Wade did, that "[c]riminal sanctions do not deter homosexual sodomy—because 'sex, next to hunger and thirst, is the most powerful drive that human beings experience,' and it is unrealistic to think that laws [proscribing sodomy] will force total abstinence."[19]

If criminal sanctions are not a deterrent, administrative sanctions may be even less effective in deterring homosexual acts within the military. There is evidence that sex—one of "the most powerful drive[s] that human beings experience"—may not be deterred even by the potential for life-threatening consequences. The director of a Gay Men's Health's Crisis has stated simply, "no one can live with 'don't' for a lifetime, especially when it comes to sex."[20] One commentator noted, "the Government's campaigns have been very effective in raising people's awareness [about AIDS] but not in changing their behaviour . . . It is difficult to legislate for behavioural patterns."[21] Indeed, in one Justice's view, to leave a person with "no real choice but a life without any personal intimacy" may violate the Eighth Amendment.[22]

Historically, this failure of deterrence has led some to advocate simply repealing the "failed" law.[23] Thus, it is not surprising those who advocate revoking the homosexual exclusion policy also advocate at least partial decriminalization of sodomy.[24] This circuitous analysis is based on the notion people are "going to break the rules anyway."[25] Of course, this reasoning is no reasoning at all. There are no rules that have never been broken. Indeed, many people routinely violate rules such as speed limits. Plainly, the fact that laws are broken does not demonstrate that they have no utility. Philosophically, the mere existence of law has an important pedagogical function. Practically, the mere existence of law has a restraining (even if not an absolute deterrant) effect.

The potential that sodomy laws will be disobeyed, on the other hand, does demonstrate that requiring homosexual soldiers to pledge to celibacy may, as a matter of policy, be an exercise in futility. Even one dissenting judge recognized that homosexuals are involved in the "frequent violation of a law that [in his opinion] has fallen into disuse," i.e., the law proscribing sodomy.[26] In any event, all the evidence supports the conclusion that a pledge to celibacy is futile. The futility lies not in any allegation homosexual soldiers might be insincere in such a pledge, but rather in the remarkable realities of the sexual dynamic.

If, as a general policy matter, a pledge to celibacy is futile, the disclosure model fails to protect either the Army or the homosexual soldier from the risks and adverse consequences of homosexual conduct in the military. The Army bears the considerable risk of a detrimental impact on good order, discipline, and morale—and hence on combat readiness. This risk carries the burden of expensive administrative and judicial processes, including not only trial by court-martial but also the defense of federal litigation.[27] Moreover, the detriment to combat readiness, whatever the underlying misconduct, cannot be recompensed by the judicial system. As one military court observed,

> The effect and impact of criminal offenses can quickly pervade and sap the morale of any unit because of the necessary intermixture of professional mission and duties with the equally necessary social and interpersonal relationships which are part and parcel with military life and tradition.[28]

The burden of repairing the effect of criminal offenses on good order, discipline, and morale falls on commanders, who then must divert time and resources to that task.

The most critical deficiency in the disclosure model is seen in the analysis in *Beller*. There the court noted that, if the military allowed homosexuals to serve, it bears the risk that—regardless of any pledge to celibacy on their part—the policy might be understood as tacit toleration, perhaps even tacit approval, of homosexual conduct.[29] This perception has been demonstrated in other cases, such as McConnell v. Anderson, where the court found plaintiff was not denied employment because of his homosexual tendencies or conduct, but because his activism regarding the societal status of homosexuals had the effect of forcing tacit approval of homosexuality upon his employer.[30]

This perception of tacit toleration or tacit approval of homosexual conduct on the part of the Army would have more than social significance. It would have legal significance as well. One could imagine, for example, that, if the Army allowed homosexuals to serve, this perception would become a basis for a turning of the tables on the concept of the status-conduct dichotomy.

Presently the argument on behalf of homosexuals is that the homosexual exclusion policy punishes homosexuals for their status and has nothing to do with their conduct. If this status-conduct dichotomy were accepted, and homosexuals were permitted to serve on that basis, their advocates might well argue that the Army had accepted homosexuals in that *status* and the Army *must have known homosexual status includes some aspect of homosexual conduct.*

This turning of the tables on the status-conduct dichotomy by homosexuals is not as unlikely as it may appear. In fact, this theory underlies arguments already accepted in Watkins v. United States Army.[31] In *Watkins*, the court held that the Army was "equitably estopped" from discharging Watkins for homosexuality because the Army "knew" Watkins was homosexual and reenlisted him in spite of that fact. *Watkins*, therefore, sowed the seeds of an "assumption of the risk" doctrine. If it lifted its ban on homosexuals serving, this doctrine—*that the Army assumed the risk of homosexual conduct by allowing homosexuals to serve*—would surely be used against the Army in an attempt to block enforcement of the Army's ban on homosexual conduct.[32] Certainly, the Army would not be able to argue that it had a rational basis upon which to conclude homosexuals would *not* commit homosexual acts over time.

Thus, the disclosure model presents considerable obstacles and risks for the Army. Nevertheless, in this model the homosexual soldier is in no more enviable position than the Army. Because his homosexuality would be known to all, his probationary status would be known to all. He might feel one misstep could pack him off to prison, yet the strain of repressing the outward manifestation of his homosexuality would be great. One former soldier discussed the strain of having her homosexuality revealed while she was in the military, and stated:

> What I did have was a massive amount of unresolved anger toward the colleague who had made the accusation [the individual who informed the Army that Humphrey was homosexual]. I lost my trust in any individual with his kind of self-righteousness. In fact, I was beginning to distrust mankind in general. I sought psychiatric help to resolve these negative feelings.[33]

This former soldier admits to lying to enter the military[34] and that the "accusation" she was homosexual was a truthful statement. Nevertheless, the trauma of losing one's career, and perhaps one's privacy, is enormous. Even if homosexuals served openly in the military, the same strain and trauma might attend revelations regarding commission of homosexual acts.

Even if the eventual result of this tension were not prohibited homosexual conduct, the stress factors inherent in this model might well cause inappropriate coping behaviors or an exacerbated sense of disenfranchisement or alienation. These feelings of disenfranchisement and alienation might result in security or disciplinary problems—and they certainly would not be positive experiences for homosexuals. Indeed, some military members who were court-martialed for homosexual conduct saw themselves as *victims* who were punished—not for their criminal acts—but for "being homosexual."[35]

Those former service members who express bitterness about the military's discovery of their homosexuality had often entered the military through deceit and with full knowledge that their homosexuality disqualified them for military service. One former service member, for example, stated she knew she was homosexual, but enlisted by saying "what [the military] wanted to hear."[36] When her homosexuality was discovered, she was discharged.[37] She describes her feelings about this incident as,

I pay my taxes because I don't want to go to jail, but I'd never do anything service-oriented for this country, and I'm not a patriot. You know, I figure that I gave them all I had so very long ago and they just f----d me over. So f--- them now![38]

A former airman claimed he "f----d everything that moved" while he was in the military, but he was embittered over his discharge because he felt the Air Force "never proved anything."[39] A former captain and squadron commander stated,

I confessed to three charges of sodomy, two counts of fraternization, and one count of official false statements, which was that "I'm not a homosexual" statement. There was plenty that [the Office of Special Investigations] didn't know. . . . A lot of people attending the trial were my troops, people who wanted to see me get what they felt was my just deserts. . . . I was known as a kick-ass commander, and so the troops that I had given reprimands to were there to watch me suffer. . . . I spent nine months [in] confinement . . . For being gay![40]

Disclosure would not "immunize" soldiers who commit homosexual acts or engage in other disruptive or discreditable homosexual conduct from court-martial charges or adverse administrative action. Thus, this model does nothing to ameliorate the inherent stress between homosexuality and military service.

Finally, the homosexual soldier's known vulnerability to prosecution if he fails—or is perceived as failing—in his pledge to celibacy could become a divisive element or basis for manipulation or coercion among soldiers. Accusations of homosexual conduct—sincere or false—could be prompted or threatened based on knowledge of another's homosexuality. In one military case, for example, the facts were that the First Sergeant was known in the unit as the "First

Skirt" and he had been previously investigated for homosexuality. The accused was charged with false swearing after he alleged the First Sergeant made homosexual advances toward him.[41] In another military case, the accused claimed he participated in drug activity because his commander had pressured him into performing homosexual acts.[42]

Where the subject of the accusation is known to be homosexual, the homosexual, as well as the accuser, might feel such accusations, or threatened accusations, would have enhanced credibility based on the fact of homosexuality.

In sum, the disclosure model fails to address the problems of homosexuality in the military. Rather, it raises serious problems for both the Army and the homosexual soldier. It is not a better solution for either the Army or the individual than the solution effected by the homosexual exclusion policy.

THE "DECRIMINALIZATION" MODEL:
Amendment of the Law Proscribing Sodomy

The "decriminalization model," similarly to the "disclosure model," assumes that controlling *criminal* conduct is the sole rationale for the homosexual exclusion policy. Thus, the model solution is to *decriminalize* sodomy. The proposal is to repeal the law proscribing sodomy or to limit its scope to acts that occur "on duty, in the barracks, on board ship or aircraft, in a situation that would create the appearance or prospect of favoritism within a chain of command, or in a situation that otherwise causes actual prejudice to good order and discipline."[43]

In military law, the proscription of sodomy is found in Article 125 of the Uniform Code of Military Justice.[44] The offense of sodomy has a single element of proof: that "the accused engaged in unnatural carnal copulation with a certain other person or with an animal." Elements in aggravation are "with a child under the age of 16" and "by force."[45]

The decriminalization model proposes amending Article 125 by adding two elements of proof, one as to whether the crime occurred on or off duty, the other as to the "actual" prejudicial impact of the crime.[46] These elements are reminiscent of the service-connected jurisdictional analysis that was overruled by the Supreme Court in Solorio v. United States.[47]

The decriminalization model has several serious flaws. If, as has been shown, criminal sanctions have little deterrent effect on homosexual conduct, expanding the elements of proof may serve only to diminish further the impact of the sanction. Indeed, the situations this new Article 125 would *not* cover far outnumber those in which it would be applicable.[48] At the same time, expanding the elements of proof would increase the difficulty of prosecuting conduct that plaintiffs, courts, and legislative drafters repeatedly have recognized as plainly prejudicial to good order, discipline, and morale. Neither result sensibly furthers the model's objective to ameliorate the problems of homosexuality within the military without excluding homosexuals.

The most serious deficiency in the decriminalization model, however, is a practical one—the practicalities of fact and logic. The model creates anomalies and ignores the realities of homosexual conduct. Under the model, many instances of homosexual conduct (as noted in Chapter Five) would fall outside the scope of the proposed proscription, yet still be acutely detrimental to the Army or even to the individual soldier.[49] Thus, it is anomalous, as well as unworkable, to decree homosexual acts are lawful in some instances but not in others.

Moreover, it is unclear why homosexual sexual offenses should be proscribed only in limited situations while other types of sexual offenses are not similarly limited to certain locations or participants or by "actual" results.[50] Clearly, whether or not conduct is prejudicial to good order and discipline does not depend upon the relationship between the actor and the military, but on the effect of the conduct upon the service.[51] For example, a military court held a sailor running a prostitution ring exposed the military community to disease.[52] There is no principled reason why this conduct could be held prejudicial when, under the decriminalization model, many scenarios of homosexual "cruising" would be permissible.

Nevertheless, one commentator states, "the real problem for the military is not the service member who engages in sexual activity on his or her own time, away from the military installation or vessel."[53] In other words, in the commentator's view, the only problem for the military is homosexuals who engage in sexual activity on Army time and on-post or aboard ship. This analysis, as mentioned above, is strikingly similiar to a jurisdictional analysis already rejected by the Supreme Court.[54] This analysis also was rejected by implication in *Hardwick*, where the Supreme Court explained,

> if respondent's [challenge to the law proscribing sodomy] is limited to the voluntary sexual conduct between consenting adults, it would be difficult, except by fiat, to limit the claimed right to homosexual conduct while leaving exposed to prosecution adultery, incest, and other sexual crimes even though they are committed in the home. We are unwilling to start down that road.[55]

Applying the Supreme Court's analysis to this proposal, the question is: Why not limit the scope of adultery, incest, and other sexual crimes to those committed "on duty, in the barracks, on board ship or aircraft, in a situation that . . . otherwise causes actual prejudice to good order and discipline"? Since no principled answer to this question exists, the underlying proposal cannot withstand close analysis.

Finally, the decriminalization model is deficient because it ignores the practical realities of homosexual conduct. The model's premise is that homosexuality *would be* compatible with military service *but for* the fact homosexual acts are criminalized. This simplification of the rationale for the homosexual exclusion policy is not supported by the law or by the evidence. Thus, this model fails to account for the whole complex of factors addressed by the homosexual exclusion policy. This model, therefore, is not a better solution than that effected by the present policy.

THE "ACCOMMODATION" MODEL

In the accommodation model, the Army attempts to adapt to the unique needs and circumstances of homosexuals through education, training, and changing the way the Army does business. This model is based on the fact that some homosexuals enter the military in spite of the exclusion policy.[56] The model also is based on (and purportedly strengthened by) the huge numbers of homosexuals some commentators believe are serving covertly in the military at present.

Most commentators who believe sheer numbers support opposition to the homosexual exclusion policy rely heavily on Kinsey's estimate of the incidence of homosexuality within the general population to estimate the numbers of homosexuals covertly in the military.[57] There are three problems with using the supposed numbers of homosexuals in the military as an analytical ground.

First, Kinsey's data on the incidence of homosexuality in the general population is increasingly questioned.[58] One recent study charges that:

> the Kinsey data is skewed toward aberrant sexual behavior because 25 percent of the male sample had a prison or sex-offender history. At least 200 of the subjects were male prostitutes. . . . Kinsey's research included "several hundred" youths ages 2 months to 15 years old . . . [and] masturbation of infants by "technically trained" experts.[59]

Reports indicate "current data show that the homosexual population is only 1% to 2% [of the general population, rather than at least 10% as asserted by Kinsey]."[60] For example, a medical study showed 1% of college students described their sexual orientation as homosexual.[61]

The second problem with relying on Kinsey's data to estimate the number of homosexuals in the military is that, regardless of the actual statistical incidence of homosexuality in the general population, there is no scientific methodology by which to extrapolate that statistical incidence to the military population.[62] The obvious treachery of this logic can be seen in *Baker*, where the court found, based on Kinsey's data, that 5% of the general population was homosexual, therefore 5% of Texans were homosexual, therefore there were "at least 500,000 exclusive homosexual males in Texas."[63]

There is, however, no scientific or common sense basis upon which to conclude that *any* sub-group of the general population is proportionately represented within the military population. Statistics on the rates at which women and racial minority members join the military prove the fallacy of such an approach.

The third and most lethal problem with using estimates—any estimate—of homosexuals in the general population or in the military as a basis for analysis is that *the statistical incidence of homosexuality simply is not relevant to whether or not homosexuality is compatible with military service.* Ten percent of Americans may have flat feet, but that fact does not demonstrate that 10% of America's soldiers have flat feet, nor does it demonstrate that flat feet are compatible with military service. Therefore, any discussion of the size of the homosexual population misses the mark in terms of evaluating the homosexual exclusion policy.

Regardless of the numbers contemplated by the model, the model's premise is, since some covert homosexuals presently are in the military, "it cannot tenably be argued that homosexuals prevent the military from accomplishing its mission."[64] The proposal, therefore, is "[i]f homosexuals are going to be in the military regardless of all efforts to keep them out . . . the military should adjust to that reality" and accommodate them fully.[65] Given the prevalence of this argument, obviously it is necessary to examine this model for logical soundness and practicability in the military setting.

LOGICAL SOUNDNESS OF THE ACCOMMODATION MODEL FOR HOMOSEXUALITY IN THE MILITARY

The logical soundness of the accommodation model depends on the strength of the model's premise and its factual predicates. The accommodation model relies on the premise that if a law is violated, it should be abandoned. Stated another way, the premise is that if the Army is unsuccessful in excluding some homosexuals, it should therefore allow all homosexuals to serve.[66] In a broader perspective, this premise raises the question of whether the function of the law is to reflect reality or rather to reflect the reality society is striving to achieve through the rule of law. Logically, law must be the latter. Thus, the fact that some homosexuals evade the exclusion policy does not demonstrate that homosexuality is compatible with military service. This analysis also holds true in regard to all other groups that are subject to exclusion policies, and even to individuals who are not eligible for service.

A second logical weakness in the accommodation model is its factual predicate. The model equates an army where some homosexuals serve covertly by evading the exclusion policy with an army where all homosexuals are allowed to serve openly. This equation is completely indefensible. It simply is not credible to extrapolate from an army where some break the rules to an army where the rules are removed.

Clearly, the differences between these two armies would be both qualitative and quantitative. For example, increased numbers of homosexuals might be attracted to the military if they could enter the service without deception and then serve openly. This phenomenon has precedent. In *Dillard*, a challenge to the single parent exclusion policy, the court noted that "[f]or a period of time . . . prior to Dillard's enlistment, single parents . . . had been allowed to enlist in the [National] Guard. . . . [T]here was a *significant increase* in the enlistment of single applicants with minor children. Also, during this period, sole parents presented many problems in other personnel management areas."[67]

Increased numbers of homosexuals—in circumstances where they openly could display their homosexuality and openly enjoy the company of other homosexual soldiers—would have a different impact on the military environment than homosexuals—fewer in number—who serve covertly by evading the exclusion policy. Because the accommodation model fails to account for the critical differences between an army where the homosexual exclusion policy has, at minimum, a restraining effect and an army without restraint, the model is not logically sound.

A final observation must be made on the logic of the accommo-

dation model. The standard for assessing the impact of homosexuality—or any other service-disqualifier—within the armed
forces is not, as some have posited, whether or not those homosexuals who evaded an exclusion policy have "*prevented* the military
from accomplishing its mission."[68]

The Army need not sputter to a complete halt before it has a
rational and practical basis for policy choices. As the law recognizes,
the Army need not even "assume the *risk* that the presence of
homosexuals within the service will *not compromise* the admittedly
significant government interests asserted in [the homosexual exclusion policy]."[69] Common sense dictates that "the Army [is] understandably . . . apprehensive of the prospect that [homosexual]
desire or intent would ripen into attempt or actual performance" of
homosexual acts.[70] Moreover, *any* risk homosexuality may pose to
the Army must be resolved in favor of the service.[71]

In sum, the analysis of the Supreme Court's decision in *Gilliard* is
apt here: "[t]he evidence that a few [homosexuals] were willing to
violate the law by not [abiding by the homosexual exclusion policy]
does not alter the fact that the entire program has resulted in [the
maintenance of good order, discipline and morale]."[72]

PRACTICABILITY OF THE ACCOMMODATION MODEL FOR HOMOSEXUALITY IN THE MILITARY

Introduction: the dynamics of accommodation in general

Accommodation is always an alternative to excluding certain groups
from military service. The question is: Will accommodation succeed
and how much will it cost?

Cost—in terms of money and combat-readiness—is a vitally important factor in accommodation. Not only is cost an important
aspect of practicability, but it is also a factor in the legal analysis of
the reasonableness of efforts at accommodation. In *Turner*, for example, the Supreme Court reviewed a prison regulation that—unlike
the effect of the homosexual exclusion policy on homosexuals—
impinged on inmates' constitutional rights.[73] Even though the law
has not established any constitutional rights arising from homosex-

uality, *Turner* is instructive. Discussing the availability of alternative policies, the Court stated:

This is not a "least restrictive alternative" test: prison officials do not have to set up and then shoot down every conceivable alternative method of accommodating the claimant's constitutional complaint. But if an inmate claimant can point to an alternative that *fully accommodates the prisoner's rights at de minimis cost to valid penological interests*, a court may consider that as evidence that the regulation does not satisfy the reasonable relationship standard.[74]

Applying *Turner* generally to the question of accommodating homosexuals in the military, accommodation would be *appropriate* only if, first, *the accommodating alternatives are fully successful*, and secondly, if *accommodation can be accomplished at "de minimis cost to valid [military] interests."*

In the military context, accommodation would be *successful* if the accommodating measures permitted an identifiable group of individuals to be full participants in soldiering, as that norm is held within the military community. The corollary of this norm is, after accommodation, that the identifiable group enjoy status and respect as full-fledged soldiers within the military community.

An example that illustrates the dynamics of accommodation in the military environment is handicapped individuals. The Army could accommodate handicapped individuals by building special facilities and transportation assets, procuring expanded medical resources, modifying military training requirements, creating tailored assignment and deployment policies, and by exempting handicapped individuals from certain routine duties, such as posting guard or serving as Charge of Quarters.[75]

These measures, however, would not result in handicapped individuals being full participants in soldiering, as that norm is held within the military community. The opposite is true. Indeed, the more extensive the measures required to accommodate a group, the more likely the group will be viewed by other soldiers as essentially civilians in essentially civilian jobs, who are not subjected to the rigors of military life, but who happen to wear military uniforms.

This concept—that wearing a uniform or even being called a soldier does not mean one *is* a soldier—may not be readily apprehended by those outside the military. A civilian co-counsel once asked, "Why can't a person in a wheelchair be in the Army? Couldn't a person in a wheelchair be a cook or a clerk in the Army?" The answer is, yes, a person in a wheelchair could be a cook or a clerk in the Army. But that person could not be a *soldier*. Cooking or clerking may be a particular soldier's job, but it is not all there is to soldiering. Similar questions have been posed in reference to homosexuals. A representative of the American Bar Association once asked, "Why couldn't homosexuals be in the Army if they were in non-sensitive positions?" The answer is—for one thing—there are no truly non-sensitive positions in the Army. For another thing, there are no "lone" soldiers. As has been recognized, a soldier's conduct—homosexual or otherwise—"at the very least has an impact upon other soldiers."[76]

The perception of a group as only "civilians in uniform" becomes especially acute as the accommodation of the group extends closer to precluding their deployment to combat theaters. General John W. Foss, Commander, U.S. Army Training and Doctrine Command, has stated the "primary mission of projecting land combat power will place a premium on the *deployability* of our Army."[77] If deployability is key for the Army, a group that cannot deploy cannot be full-fledged soldiers—regardless of the reason for non-deployability. This could result in a perception of the group as "second-class soldiers." Indeed, this is one danger inherent in legislative initiatives during the Gulf War to accommodate single parents or dual-soldier married couples by precluding deployment to combat theaters. If homosexuals were not fully deployable—for example, because of issues of privacy, close quarters, host nation laws, and so on—the prospect of creating a group of second-class soldiers indicates accommodation of homosexuals in the military would be less than successful.

The example of handicapped individuals illustrates how some groups, because of their general circumstances, simply are not able to achieve status as full soldiers within the military community. This is not an example of "illegitimate bias" against handicapped people. Rather, it is an example of military social and operational norms. Thus, even apart from issues of the administrative burden and cost-effectiveness, accommodation poses an intrinsic—although not always insurmountable—danger to good order, discipline, and morale. Accommodation raises the grave prospect of an army ultimately divided between those who can be called upon to suffer the mass and minimally private living conditions of an army in the field and those who are excused from that requirement. This danger of an army divided, and related systemic effects on unit cohesion, must be addressed forthrightly if accommodation— regardless of the group accommodated—is to profit the Army and the individuals involved.

The proposed accommodation model: focus

The accommodation model focuses primarily on two regulatory rationales for the homosexual exclusion policy, maintaining "the integrity of the system of rank and command" and facilitating the "assignment and worldwide deployment of service members who frequently must live and work under close conditions affording minimal privacy."[78] The model does not further address the problem of homosexual conduct within the military. As is usual by definition in any model of accommodation, more emphasis is placed on changing the accommodating community than on changing the accommodated group. This section will discuss the impact of accommodation in the context of the regulatory rationales for the homosexual exclusion policy, as well as the systemic effects of accommodation.

Impact of accommodating homosexuality on the integrity of the system of rank and command. The integrity of the system of rank and command obviously is a crucial component of combat readiness. In the accommodation model, the military's concern has been expressed as "[t]he fear that openly homosexual supervisors could not command respect."[79]

The potential that a homosexual may be unable to command respect as a soldier is not a "fear" on the part of the military, but a

fact of life. In *Woodward*, for example, the court noted that the plaintiff, who was discharged for homosexuality, informed the Chief of
Naval Personnel: "I am well aware of the problems of social acceptance and special problems of leadership with which I will be confronted as my associates become aware of my homosexuality."[80]

On the issue of whether "openly homosexual supervisors could
. . . command respect," Judge Norris' dissent in *Miller* is enlightening. He states:

> Much of the tension and hostility, absence of trust and respect [and]
> offended sensibilities, . . . flows not from the fact of a private sexual
> act, but rather from related public behavior. *Yet much potentially offen-
> sive public behavior other than actual sexual activity is beyond the [military's]
> power to control. Public expressions and encouragements of homosexuality as
> personal philosophy, and public displays of behavioral tendencies or traits
> commonly perceived as homosexual, clearly are more likely than private
> conduct to arouse hostility, create tension, or offend sensibilities. Yet such
> disruptive public behavior may not constitutionally be prohibited by the
> [military]*.[81]

It was absolutely correct to point out that many types of homosexual public behavior may be as disruptive or even more disruptive
than actual homosexual sexual activity. In *Smith*, for example, a
homosexual sued on the claim he was discriminated against for
being too effeminate—his conduct was disruptive but not prohibited.[82] The conclusions the dissent drew, however, are not well-
founded.

First, it is not true the military must tolerate any form of disruptive behavior. Conduct—*any* conduct—prejudicial to good order
and discipline or of a nature to bring discredit upon the armed
forces is a crime under the Uniform Code of Military Justice.[83]
Moreover, the Army has an extensive administrative process by
which to identify and discharge soldiers whose conduct—whether
or not that conduct is criminal—proves them unsuitable for military
service. This lack of tolerance for disruptive behavior is seen even in
civilian settings, for example, when one homosexual's application
for a license to enter into a homosexual marriage generated so much

disruptive publicity that the court held it was a permissible factor in the employer's decision to deny employment.[84]

Assuming for the purposes of argument, however, that the dissent was correct in finding "much potentially offensive public behavior other than actual sexual activity is beyond the [military's] power to control," it does not follow that the homosexual exclusion policy is ineffectual and therefore should be repealed.[85] A more reasoned conclusion is that *if such disruptive conduct cannot be prohibited, then homosexuals must be excluded*. Clearly, exclusion is the *only* method by which the military can avoid the adverse impact on good order, morale, and discipline that would result from the noted "tension and hostility, absence of trust and respect [and] offended sensibilities" caused by certain public homosexual behavior.[86]

In the accommodation model, the solution to the potential for lack of trust and respect is to provide homosexual soldiers with leadership training to facilitate their ability to command respect, and to enforce that training by rating supervisors on their leadership abilities.[87] At the onset, it should be noted that the Supreme Court rejected a similar proposal in regard to a statute excluding adults from dance halls catering to minors. In that case, respondent proposed that rather than excluding adults from dance halls, "the city [could] achieve its objective through increased supervision, education, and prosecution of those who corrupt minors"[88]—that is, through increased training and enforcement. The Court rejected the model solution. Instead it held the statute rational even though the "city's stated purposes . . . [might] be achieved in ways that are less intrusive."[89]

The results expected in this model by increased training and enforcement can be compared to the actual results in a different closed population, the population of those persons incarcerated in large prisons. While military life certainly does not mirror prison life, prison populations have been prime subjects for studying the dynamics of creating a culture among diverse individuals and the natural emergence of leaders and leadership schemes.

Prison populations are microcosmic societies. These populations spontaneously institute power structures and develop systems of rank and command. It is well-documented that inmates stratify themselves according to their crimes, *as perceived by the inmates*. In *Harper*, for example, there was evidence "inmates who are identified as or suspected of being pedophiles or homosexuals are a favorite target for violence since many incarcerated felons were sexually abused as children."[90] In *Espinoza*, there was evidence plaintiffs "received threats from other inmates due to their convictions for sex crimes involving a child and being labeled homosexual. Both men received threats of rape and assault and were labeled 'baby rapers' by their fellow inmates. Both men requested protective custody status."[91]

In this spontaneous process of instituting power structures, certain types of offenders frequently are stratified in such a way that they cannot command respect or hold positions of leadership in the system of rank and command within the population. This exclusion from leadership is unrelated to the individual's actual ability to lead, rather it is related to the individual's conduct. Further, the chance is small that this exclusion from leadership could be affected by enhancing the leadership skills of the excluded individuals.

Similarly, if the homosexual soldier is unable to command the respect of his subordinates, superiors, or the military as a whole, enhancing or facilitating that soldier's leadership skills does not solve the problem. This is because the failure of respect is unrelated to the actual ability of the homosexual soldier to lead.

This social phenomenon is seen in all populations that have power structures. For example, suppose a male battalion commander has a sexual preference for teen-age girls and he begins an affair with the fifteen-year-old daughter of a sergeant. Soon all the soldiers in the battalion know of the affair. Many of them disapprove of the commander's conduct. These soldiers will continue to obey their commander and render him appropriate military courtesies— this is their military duty and they will be punished if they fail in it. But this does not mean these soldiers *respect* their commander, that he enjoys their full confidence as a leader, or that the unit has the *esprit* needed for success. The commander's leadership skills are largely immaterial in this situation: the problem is his conduct. Courts as well as plaintiffs have recognized that attitudes toward homosexual conduct would affect the ability of homosexuals to command the respect necessary to perform supervisory duties.[92]

This discrepancy between respect and ability is not unique. Rather, it is an integral aspect of the view that respect is a fragile and coveted commodity that must be earned. Respect and ability are often closely related, but they are not routinely coextensive. Thus, to the extent that the model relies on increasing *ability* as a means by which to increase *respect*, the theory that leadership training effectively could address the problems of homosexuality in the military is infirm and the model is deficient on that ground.[93]

Impact of accommodating homosexuality on privacy entitlements and related military concerns. The accommodation model also addresses the military's determination that homosexuality hinders or complicates the "assignment and world wide deployment of service members who frequently must live and work under close conditions affording minimal privacy."[94] The model recognizes the positive effects of gender segregation in protecting individual privacy and promoting discipline, but concludes that the impact of homosexuals on the efficacy of gender segregation "appears to be a unit level management problem, not 'an assignment and worldwide deployment problem.' "[95]

Gender segregation, as already seen, serves several important goals of personal privacy, discipline, and cost-effective administration of the force. The accommodation model hypothesizes that what the Army achieves for heterosexuals through gender segregation, the commander can achieve through "management" in situations where the presumptions underlying gender segregation have broken down, namely where there exists sexual preference for the same gender.[96]

The strength of this hypothesis is tested quickly by applying it to heterosexual soldiers and asking whether—or how easily—the commander could manage privacy and disciplinary concerns if he did not have the tool of gender segregation. When the hypothesis is applied to heterosexuals who are not segregated by gender—in situations where gender segregation traditionally is appropriate—

the operation of the hypothesis becomes clear. The commander, for example, would be faced with the puzzling task of teaching, training, leading, exhorting, or somehow forcing his male and female soldiers not to manifest their sexual preferences while showering together.

In addition to somehow "managing" the manifestation of heterosexual preference—from flirting to acts—the commander would have to train his soldiers not to feel their privacy was infringed by the presence of the opposite sex in situations, such as showering, where individuals normally are segregated by gender. Through the dedication of resources and personnel, the commander might be able to control flagrant misconduct in these conditions, but he would have little control over the overall impact on the climate of his command.

Plainly, this model expects the impossible, or at least the herculean. The commander is faced with privacy problems and a potential breakdown in discipline if he integrates homosexuals with heterosexuals of the same sex under conditions affording minimal privacy. The commander is faced with different, but equally serious, privacy problems and potential indiscipline if he segregates homosexuals with other homosexuals in close living conditions. A former service member, discharged for homosexuality, puts it this way: "[t]he military fears that if the homosexual is allowed in its ranks, there will be . . . rampant sexual activity among the men and women (men with men, women with women)."[97] It cannot be said such concern is completely devoid of a rational basis.[98]

Indeed, in a study of heterosexual activity, "[t]wo-thirds of American troops who served in coed units during the Gulf War say men and women were having sex, and more than half said it hurt morale."[99] The study did not indicate how much sex occurred or how many individuals were involved. Thus, the military must consider the effect of heterosexual activity as well as the potential for homosexual activity on military operations.

Since the commander does not have a tool similar to gender segregation to meet the privacy and disciplinary implications of homosexuality, homosexuality—contrary to the premise of the model—does hinder or complicate "assignment and world wide deployment of service members who frequently must live and work under close conditions affording minimal privacy."

There are two possible ways to work around the practical problems homosexuality presents for managing soldier assignments and deployments. These are to change the aspects of Army life that create "close conditions affording minimal privacy" or to implement assignment or deployment limitations for homosexuals. This section will explore these alternatives in turn.

CHANGING THE WAY THE ARMY DOES BUSINESS

If homosexuality presents problems in situations where soldiers must live and work together in close conditions affording minimal privacy, one way to accommodate homosexuality within the military would be to change the conditions of Army life that afford minimal privacy. This would minimize the need to depend on gender segregation for protection of privacy and discipline.

To make this accommodation, the Army would have to change the way it does business. For example, it would have to abandon its focus on the most efficient, most cost-effective methods to get the largest number of soldiers rested, showered, dressed, and inspected in the shortest period of time. Instead it would have to refocus on the creation of a more "individualized lifestyle" by providing soldiers with private rooms, private bathroom facilities, and so on.

The Army sometimes has taken a more individualized approach to military living. But this approach may not always be possible, cost-efficient, or conducive to the military's disciplinary or operational goals. The Army may opt to establish different living arrangements at any time or be required to do so by military or operational necessity. Moreover, how soldiers are billeted at post, camp, or station does not resolve the problems homosexuality presents for assignment and deployments. The abiding fact of army life is that, in a moment, the soldier can be ordered into long-term, high-stress situations where privacy is at a premium and discipline is decisive. For the soldier, more than for anyone else, "[t]he facts of war are often in total opposition to the facts of peace."[100]

In sum, even if the Army decided it was profitable to change the way it does business in order to accommodate homosexuality, the Army can only change so much. Regardless of how soldiers live at post, camp, and station, the Army must live by the creed that "the inescapable demands of military discipline and obedience to order cannot be taught on battlefields."[101] The Army cannot change the fact that it is an *army*, and that therefore—by the very nature of fighting wars and training to fight—its soldiers will be required to "live and work in close conditions affording minimal privacy."

ASSIGNMENT AND DEPLOYMENT LIMITATIONS

The second way to address the problems homosexuality presents in "close conditions affording minimal privacy" is to impose assignment and deployment limitations. Without assignment and deployment limitations, homosexual soldiers must be deployed into situations of minimal privacy, while the commander is expected to do the best he can to maintain good order, discipline, and morale. That is, at the same time he is taking his unit into a special high-stress, high-stakes environment, the commander must also fully shoulder "the risk that the presence of homosexuals within [his unit] will not compromise" his soldiers' combat readiness.[102]

The risk of homosexual misconduct—along with the potential disruptive effect of other homosexual conduct such as flirting and displays of affection[103]—is not like the risk of other misconduct. Homosexual acts are motivated by a *drive*. Rape, for example, is widely considered to be motivated by rage or a desire to exert power over others. Rape generally is not motivated by a sexual urge that is an expression of one's sex drive. The drive for sex is basic to all humans, while feelings of rage, inferiority, or powerlessness are peculiar to some individuals. The fact that homosexual acts are the expression of a drive makes them, and the risk or likelihood of their occurrence, fundamentally different from other acts or other misconduct.

When homosexual conduct occurs in a deployment situation, this could severely disrupt the mission or even provoke an international incident. For example, in the Gulf War several service members who were caught in homosexual acts had to be returned immediately to the United States from Saudi Arabia.[104] Under Saudi religious law, homosexual acts are punishable by execution. Thus, immediate evacuation was required to avoid having these incidents come to the attention of the Saudi police.

Plainly, the risk of homosexual conduct and its potential to disrupt operations or even alliances is not a "unit level management problem." Imposing this risk on unit commanders does nothing to ameliorate the problems presented by homosexuality in the military environment. Thus, assignment and deployment limitations must be imposed.

Assignment and deployment limitations, as discussed above, risk divisiveness by creating a group that is perceived as less than full-fledged soldiers because policies limit their service to "easy" stations and "clean" duties. There are always some soldiers who manage extended periods of service in garrison or at posts with the nicest facilities or best locations. This is perceived by other soldiers essentially as a matter of good fortune. One soldier's good fortune, however, has a different social connotation than the situation where an identifiable *group* of soldiers consistently is assigned to what may be perceived as desirable stations or duties—for example, positions where the soldier never has to go "to the field" (a synonym in the soldier's mind for being tired, dirty, hungry, and either dripping with sweat or shaking with cold)—as the result of official policy.

The effect of assignment and deployment limitations for homosexuals may be divisive in other ways. Unlike soldiers who are simply lucky in the duties or stations they draw, duty limitations for homosexuals would be based on *behavior* that is disruptive of the military mission. The logic of behavioral-based assignment limitations might be difficult for other soldiers to accept. Clearly, for the most part of the army day, soldiers are being told to and trained how to *change* their behavior to match that required for successful soldiering.

Soldiers might conclude, therefore, that deployment limitations "afforded special status to a behavior-based group, homosexuals."[105] Soldiers might conclude further that homosexual behavior is an inappropriate ground for what is perceived as preferred treatment in assignments or duties. To the extent soldiers perceive the accommodation of homosexuals as unfair, the model creates or exacerbates tensions between the accommodating community and the accommodated group.

The accommodation model raises the specter of unfairness not only in the sense of the perception of special treatment for homosexuals, but also in the absence of special treatment for other groups within the larger military community. If one group, homosexuals, is accommodated because of its behavior—which in this case carries the potential for criminal behavior—other groups rightly may demand accommodation. For example, single parents—a group whose characteristic behavior relates to an understandable and laudable desire to provide adequate care for their children—could demand similar special treatment in assignments and duties. Thus begins the division of the Army into "first" and "second class" soldiers.

This predictable phenomenon was seen in Jones on Behalf of Michele v. Board of Education.[106] In *Jones*, female students sued to prevent gender integration at their all-girl school. Their claim was based partially on the fact the New York City Board of Education had "identified homosexuals as needing special educational assistance" and the Board had established a special school for homosexual students. The girls sued for a special school for girls. The court found, however, *"the mere fact the Board, in the exercise of its educational discretion, has identified homosexuals as requiring special treatment does not give rise to any obligation to do the same for girls."*[107]

The court in *Jones* further found that girls could not be regarded as similarly situated with homosexuals—and thus be given a special school—because girls were nearly a majority of the student population, while homosexuals were a much smaller percentage.[108] The court held, therefore, *because girls were in the majority, the girls did not have an "equal right to be treated differently."*[109] Needless to say, the Army need not court the phenomenon of each group within the military community seeking special accommodation.

Neither changing the way the Army does business, nor imposing assignment and deployment limitations for homosexuals results in successful accommodation of homosexuality within the armed forces. Both courses of action present risks to the Army and the individual. A policy-maker rationally could conclude these risks are better avoided by the homosexual exclusion policy.

Systemic Effects of Accommodating Homosexuality in the Military

The accommodation model, as seen, affects the system of rank and command and the assignment and deployment of soldiers. Accommodation also has systemic effects on the military. Successful accommodation of homosexuality in the military necessarily would require successful integration of some important features of homosexual culture, as well as adaptation of the accommodating culture to the special needs and requirements of homosexuality. This section will discuss some of the potential stress points between homosexual culture and military culture.

Cultural integration A culture is defined by its language, dress, traditions, and social norms. The military has an extremely well-defined culture. So does the homosexual community. One feature of military culture traditionally recognized as important to soldiering is the availability of religious resources, such as chaplains. As recognized in Katcoff v. Marsh, the chaplaincy program is relevant to and reasonably necessary for the Army's conduct of national defense.[110] Accommodation of homosexuality in the military would implicate this aspect of military culture in two ways, both with the potential for an extremely adverse impact on the Army's ability to dedicate its attention and resources to military matters.

First, homosexual cultural movements, for a variety of reasons, are tied to profound feelings about religion and religious issues. One report in a national magazine showed, for example:

In the past 18 months or so, many churches across the country (six in Los Angeles alone) have been vandalized and broken into, with gay activists claiming responsibility. . . . [This is] a straight-out hate campaign. A catalog to an AIDS art show . . . reflects the general tone: Cardinal O'Connor is a "fat cannibal in skirts," and his cathedral is "a house of walking swastikas." . . . Savage mockery of Chris-

tianity is now a conventional part of the public gay culture. . . .
[O]ne newly ordained young priest and his elderly mother [were
pelted] with condoms until police intervened and escorted them
away.[111]

Secondly, although homosexuality is accepted in some religions
or denominations—Reformed Judaism, Reconstructionist Judaism,
and the Unitarian Universalist Association are reported to have
national policies permitting ordination of active homosexuals[112]—
homosexuals increasingly have turned to specialized congregations
for spiritual support.[113]

To accommodate homosexuality in the military, homosexuals
similarly may conclude that the military chaplaincy must accommodate homosexual parishes as well as homosexual clergy. Indeed,
under an accommodation model, homosexual clergy would not be
disqualified, by reason of homosexuality, from appointments as
military chaplains at large. Whether homosexuals should be ordained or not—or whether organized religion is responding rightly
or wrongly to the needs and problems of homosexuals—is hardly
the point. The debate is bitter on all sides, even within religious
denominations:

> some in the church believe that all homosexual activity is immoral . . .
> others find gay and lesbian sex acceptable "when practiced in a context
> of human caring and covenantal faithfulness." . . . The proposed
> [doctrinal] changes . . . are certain to touch off fierce debate in [an]
> often-polarized denomination.[114]

The point is, if homosexuality were accommodated in the military, the bitter divisiveness that attends the larger debate on homosexuals and religion could result within—or be projected onto—the
military community. Suffice it to say, for military, social, and constitutional reasons, the Army should avoid anything that might em-
broil it in this debate.

Besides religion, other important concerns presently at issue in
homosexual culture regard homosexual marriage and adoption of
children by homosexuals. The military already has addressed

issues created by a soldier's marriage to a female-to-male trans-sexual.[115] To accommodate homosexuality within its ranks, the military might be required to grapple with other "family" issues as well. The military sometimes bases benefits on marital or parental status. These policies might be challenged by homosexual soldiers on the same grounds homosexuals challenge local and state benefit programs and federal laws. For example, in Minneapolis, three library employees were awarded more than $90,000 in damages when an administrative panel ruled the city's failure to provide health insurance for their lesbian partners was "discriminatory."[116]

Homosexuals have already claimed that a "fundamental right to marry" was burdened by an immigration statute.[117] Another homosexual attempted to obtain a visa as "an alleged spouse of an American citizen."[118] A homosexual attempted to inherit another homosexual's estate on the ground "his [plaintiff's] constitutional rights had been violated because he and [decedent] were prevented from being legally married because of the state prohibition against same-sex marriages."[119]

Moreover, what might seem like a logical check on such claims by homosexual soldiers is not: the law actually provides a basis upon which a homosexual soldier could argue he had a marital-type relationship with a person of the same sex—and thus was entitled, for example, to a basic allowance for quarters at the married rate—*without admitting to homosexual acts.*[120] This was the precise situation in *Pruitt.* In that case, plaintiff admitted to being twice "married" to another woman, but the court declined to reach the issue of whether—by reason of the law and logic's view of marriage as a con-summated relationship—plaintiff had committed homosexual acts.

Besides policies implicating marital status, Department of Defense policy is to facilitate the adoption of children. This policy also is subject to legal challenge if it excluded homosexual soldiers as eligible adoptive parents. Indeed, *any* policy differences between the treatment of homosexual and heterosexual soldiers—regardless of legal merit—would raise the specter of litigation.

Litigation, regardless of result, is expensive and not a favored use of taxpayer funds. More importantly, as the Supreme Court has recognized, the mere pendency of lawsuits is disruptive to the military mission.[121] Precedents for the types of suits homosexual soldiers might bring already are set, however. In one case, for example, a homosexual brought suit against Disneyland for banning same-sex dancing at its establishments.[122] Another plaintiff claimed he was censured for wearing makeup and openly discussing his homosexual sexual life while heterosexuals who discussed their sex lives were not censured.[123] A homosexual male who enjoyed cross-dressing might bring suit when this conduct was found to constitute an offense under the Uniform Code of Military Justice.[124]

Denials of security clearances based on homosexuality could engender repeated and protracted litigation, just as in the civilian sector.[125] This type of litigation may present unique and substantial policy burdens. For example, in one case where the Department of Defense declined to process a security clearance application based on plaintiff's homosexuality, the agency was required to submit classified documents to the court for in camera inspection for purposes of establishing the state secrets privilege.[126] Moreover, homosexuals surely would argue that "the Army allowed me to soldier, therefore the Army must allow me a clearance." Thus, a new variation on the assumption of risk doctrine would be born.[127]

Therefore, to accommodate homosexuals, the Army would have to change security clearance procedures—that is, deem homosexuality irrelevant to the security inquiry—or accept the increased risk of burdensome administrative and judicial challenges to the clearance process. This analysis may apply across the board to military personnel policies. It is not unforeseeable that the cadre of homosexual rights groups and legal defense funds that supported plaintiffs challenging the homosexual exclusion policy would continue to seek to further their cause within the military if homosexuals were permitted to serve. Indeed, homosexual rights groups, have already announced if the ban on homosexuals is lifted they will challenge the military's policy on HIV testing. A spokesman for one group states the HIV policy is discriminatory because it would disqualify more homosexuals than heterosexuals for military service.[128]

Other features of homosexual culture that might impact on whether or not homosexuality could be accommodated successfully in military culture are homosexual bars and organizations or publications oriented to homosexuals. The homosexual bar is a feature of homosexual culture that reflects the desire of homosexuals to socialize together.[129] Homosexual bars rarely are integrated with heterosexuals. An establishment usually becomes a "gay bar" by establishing a homosexual-only patronage. The practical effect is the exclusion of heterosexuals. A similar social pattern could result in military establishments, such as recreation centers or officers' clubs. If the establishment was patronized primarily by homosexual soldiers, the practical effect of the exclusion also might extend to homosexual soldiers who did not desire to be part of the "open" homosexual community.[130]

A comparable phenomenon might occur if homosexual soldiers gravitated toward certain installations or were concentrated in certain military occupations. These installations or military occupations would become known for their numbers of homosexuals and the influence of homosexuality as a sub-culture. This in itself would have a unique and not entirely positive impact on the accommodation of homosexuality within the larger military culture.

If homosexuals are accommodated within the military, then it is incumbent on the Army to determine, for example, whether or not soldiers can buy homosexually oriented magazines at the post exchange, organize homosexually oriented social clubs, athletic teams or events on post, or establish other culturally-based adjustments within the larger military culture. This process of cultural adjustment and its potential effect on social institutions has been seen in other settings. For example, in a discussion on addressing the concerns of homosexuals in a university setting,

[d]iscussion has included the possibility of creating a gay studies major, extending bereavement leave and health benefits to the partners of gay employees, expanding student housing to include gay

couples, designating special dorms as "safe residential space," creating a center or lounge for gay students and offering more lectures and cultural events with gay themes. . . . The university has also . . . directed the library to expand its acquisitions of books relating to homosexuality.[131]

Measures such as those discussed above implicate a wide range of social norms deeply rooted in the historical development of Western culture. Far-ranging departures from social norms predictably provoke dispute, even when pursued in a legislative setting where individuals have a semblance of control over their destiny through the representative process. Thus, far-ranging departures from social norms have an even greater prospect for social disruption when pursued in a closed setting ruled essentially by an oligarchy. Some students affected by the above university's desire to accommodate homosexuality,

made it clear that they did not want a portion of their tuition spent on programs and activities for homosexuals. . . . "[Y]ou have to have both sides presented and then let students make their minds up about what they believe," said Jack O'Kane, . . . who has organized the Rutgers University Heterosexual Alliance. . . . "Right now you guys have a monopoly on this issue."[132]

The effects of accommodating homosexuals in social institutions may vary. Students in the Rutgers University Heterosexual Alliance believed that the homosexual agenda was receiving disproportional support at their university, just as the students at the all-girl school in *Jones* believed that the school board gave disproportional support to the all-homosexual school.[133] This belief has been expressed by members of other social institutions. For example, one religious commentator states:

There is another element—an important one—in the debate over gay clergy . . . expressed from East Coast to West. Of those theological schools that have been most open to gays, many have found that gay faculty and students have in turn attempted to influence curriculum, faculty appointments, and other matters. Given all the other difficulties that seminaries face, the possibility of a "*gay veto*" is not an issue that most schools are eager to take on.[134]

Thus, even if the homosexuality of individual soldiers did not detract from successful service, the Army might constantly be faced with military, political, social, and moral issues regarding homosexuality. These issues could range from federal litigation at the departmental level to a request to the post commander to advertise a homosexual event in the installation newspaper. In *Mississippi Gay Alliance*, for example, the court refused to compel a campus newspaper to run a paid advertisement for a homosexual group. The court noted that homosexual conduct was a crime and that "newspapers have the right to avoid becoming even peripherally involved in criminally related activity."[135] But in Gay Alliance v. Matthews, the court allowed the group to advertise because it found there was no claim or evidence that it participated in criminal homosexual acts.[136]

Clearly, actions that may be perceived as sponsoring homosexuality, or as condoning a sub-culture based on prohibited or disruptive conduct, might have negative repercussions within the military culture. Even the amount of time, effort, and expense dedicated to the attempt to sort out whether an act sponsors or condones homosexuality might have such a negative impact. Certainly, the "calm, serenity, and aloofness from controversy" the Army requires to concentrate on its military mission would be jeopardized by the myriad issues presented during attempts to integrate homosexual culture into military culture.

Like the homosexual exclusion policy itself, these myriad issues of cultural accommodation must be reviewed in their military context. The more time and effort the Army, its commanders, and its non-commissioned officers must direct toward resolution of practical problems and social and political issues, the more time and effort is subtracted from military training—training that is even more critical as the force scales down in the future. As the Army's Chief of Staff has observed,

[W]e have begun to shape a smaller Army . . . So every soldier, every unit, and every leader within our smaller force structure must be fully

trained to fight and win. . . . [w]e must train with our eyes firmly
fixed on our sacred responsibilities to the sons and daughters of this
nation who are entrusted to our care. . . . [W]henever a sergeant takes
the extra time to plan his training in precise detail, whenever he
spends those extra hours executing his training to exacting standards,
whenever he devotes that extra effort to scrupulously assessing his
training, *he is investing in the lives of his soldiers*.[137]

Plainly, the Army must not lightly invite training detractors—
whether social or political controversy or labor-intensive practical
problems—into its ranks.

Community adaptation Besides cultural integration, the accommodation of homosexuality within the military would require community adaptation. This adaptation might include the necessity for
special facilities, additional personnel (especially for the provision
of medical care), and special living arrangements. As previously
discussed, homosexuals may require specialized medical resources to meet their sometimes unique physical and emotional
needs.[138] And accommodation poses other practical adaptation
questions, such as whether or not homosexual soldiers should be
allowed to room together in the barracks or share government
quarters.

From the military point of view, just the fact of two homosexual
soldiers living together might be deleterious to good order, discipline, and morale. Certainly, no commander would authorize heterosexual male and female soldiers to room together in the barracks,
regardless of how sincerely they disavowed any sexual intent toward one another. This point again raises the practical problems
addressed by gender segregation, which cannot be solved in the
same way in regard to homosexuals. Yet if homosexuals are treated
differently, there is the danger that they will be viewed by others as
receiving "preferential treatment" or that the matter will become
grounds for expensive litigation, or both.

Special living arrangements for homosexuals, however they
might be appropriate, would require the Army to make policy
choices that might cause contention or result in litigation, no matter

what policy choice the Army made. For example, if the Army pro-hibited homosexual soldiers from living together in the barracks, this restriction would be challenged. If homosexual soldiers were allowed to live together, heterosexual soldiers would challenge rules prohibiting them from co-habitating in the barracks.

If the Army allowed homosexuals to live together and then had a legitimate occasion to inquire into the arrangement, this policy would be challenged. Moreover, so long as sodomy is a crime, homosexual soldiers could not have a claim to special living arrangements—in other words, rooming with other homosexuals and so on—if they admitted homosexual acts were in issue. In *Aumiller*, for example, a homosexual university professor living with a homosexual student obtained a "protective order restricting [the university's] ability to inquire into [his] living arrange-ments."[139] Thus, to meet a homosexual soldier's challenge to the Army's legitimate inquiry into these situations, the Army would be required to investigate soldiers' private lives or accept the premise that these living arrangements in all instances were purely platonic.

Special living arrangements—or special clubs, social facilities, and so on—also risk reinforcing a behavior-based group, homosex-uals, as a sub-culture within the larger military community. Thus, special arrangements might be necessary for community adapta-tion to the accommodation of homosexuality within the military, yet they might actually work against accommodation at the same time.

This phenomenon already is observed in military culture when soldiers are grouped—in social not offical terms—by behavior, or potential behavior, that is considered substandard for soldiering. These groups of soldiers frequently become a focal point for rein-forcing the military identity of the larger group. One former Army officer notes his former "unit had gotten a nickname throughout Germany as the 'Gay Berets' as a consequence of my court-martial [for homosexual sodomy]."[140] Groups of soldiers with a reputation for drug use may be called the "peyote platoon," those with a reputation for being overweight may be called the "fat boys."

Desirable or not, this social phenomenon is a reaffirmation of the larger group's identity as *"real"* soldiers. Nevertheless, this social phenomenon clearly can go too far and result in divisiveness. Controlling this natural social phenomenon would be more difficult if soldiers were grouped by non-military behavior as a matter of offical policy.

Conclusion

Social and political issues such as the legitimacy of homosexual clergy, the appropriateness of homosexual marriage, and the wisdom of homosexual adoption, as well as other issues of culturally-based adjustment, are some of the most controversial ever confronted by society as a whole. Controversy, however, is part and parcel of larger society. In fact, controversy often benefits rather than impedes the democratic process.

Similar conclusions do not apply within the military, however. Indeed, "what may be questionable behavior in civilian life, and yet not present any danger to our form of Government, may be fatal if carried on in the military community."[141] As the Supreme Court stated in Goldman v. Weinberger, "[t]he military need not encourage debate or tolerate protest to the extent that such tolerance is required of the civilian state."[142] Nor, as a corollary, must—or even should—the Army be forced to grapple with and resolve issues that are confounding society at large. From the military's point of view, "those who . . . work to protect our nation should not be required to toil in contention and strife engendered from within. It is enough that they might be required to labor while being critically assailed from without."[143]

The wisdom of the Army in avoiding issues that are confounding society at large is especially apparent since the process of confrontation and resolution of these issues—particularly to the extent it involves litigation—would require the expenditure of scarce resources, financial and otherwise, better spent on accomplishment of the military mission.

SUMMARY:
The Accommodation Model for Homosexuality
in the Military

The Supreme Court has discussed the accommodation of individual rights in other settings that have a stringent requirement for order and discipline. Applying the Court's analysis of a regulation in Turner v. Safly[144] to the military, the Court's view could be stated as follows: "When accommodation of an asserted right will have a significant 'ripple effect' on fellow [soldiers], courts should be particularly deferential to the informed discretion of [military] officials."[145] Deference is even more appropriate in evaluating the homosexual exclusion policy because, in contrast to *Turner*—where the Court reviewed the impact of a regulation on an *established right*—no constitutional rights have been established arising from homosexuality.

Regardless of any disagreement over the nature, or even desirability, of the "ripple effect" of accommodating homosexuality within the military, it is clear that the effect of accommodation would not be neutral. It is equally clear that the potential effect of accommodating homosexuality within the military is significant. The potential effect of accommodating homosexuality reaches not only to fellow soldiers, but to the parameters of military culture itself. Accommodation is not more advantageous to the Army than the present homosexual exclusion policy. Rather, accommodation of homosexuality presents significant risks of a controversial and costly adverse impact on good order, discipline, and morale. Just how serious and far-ranging these risks might be is seen in the regulatory rationales that homosexuality is incompatible with military service because of its impact—or potential impact—on the Army's ability to "recruit and retain soldiers" and to "maintain the public acceptability of military service."[146]

These concerns—recruitment, retention, and the public acceptability of military service—are each a function of the relationship between the American people and their army. The following chapter will examine the homosexual exclusion policy, and proposed changes to the policy, in the context of this singularly important relationship.

CHAPTER SEVEN

The Relationship Between the American People and the American Army

PRACTICAL REASONS FOR POLICY-MAKING RESTRAINT

*When people ask me why I went to Vietnam, I say, "I
thought you knew. You sent me."*
—Lt. Col. Wallen Summers,[1]
West Point, 1971

The American soldier is related integrally, inextricably to the American people as a whole. At various times in history, the American soldier has been treated as a poor relation. At other times, he has been highly regarded as a prominent member of the American family. Regardless, the soldier knows well his destiny is tied to the will of the people who together comprise the nation he has pledged to defend.

This relationship between the soldier and the public is not a conjectural one, but rather is one founded on the fact that, "under the Constitution, the American people through their elected representatives not only raise, support, and regulate the Army but also determine the very battlefields upon which the Army must fight."[2] Thus, when the military evaluates the possible impact of a particular policy on the public, its analysis must focus on the "importance of maintaining the bond between the American people and their soldiers . . . [because that bond is] the source of [the Army's] moral strength."[3] As one military court stated, "[a]t the heart of every successful military force are morale, discipline, and public support of the cause. An army which lacks those cannot hope to succeed . . . and the nation must necessarily fail in battle."[4]

Stated simply, the consequence of an incautious policy determination that breaks the bond between soldier and public will prove a disaster. Thus, even if recruitment, retention, and public acceptability of the military had not been expressed as rationales for the homosexual exclusion policy, the American people always are at issue when the Army formulates military personnel policies. It is the American people, ultimately, who must be persuaded of the wisdom of personnel decisions and have confidence in their efficacy.

In this sense, the dynamics of the volunteer force call for professional military as well as business judgments in formulating military personnel policies. When a policy is formulated and its expected result is reviewed, military and operational considerations are paramount. But the review process also must include a sophisticated and essentially pragmatic judgment as to whether or not the policy will "play in Peoria."

Indeed, recruiting an all-volunteer force year after year from cities and towns across the nation is big business. Millions of dollars are spent annually on advertising. Thousands of soldiers are dedicated to full-time recruiting duties. These efforts and funds are wasted, however, if military personnel policies create an environment within the force that is unacceptable—or even perceived as unacceptable—to those otherwise most likely to desire military service. The cost in terms of national defense is even greater than the mere waste of recruiting efforts and funds.

The defense needs of the nation exist apart from military recruiting success. If recruiting efforts fail to meet manpower needs because, for example, military service is poorly regarded by the general public, it may become necessary to lower military enlistment or retention standards. Lower standards, however, may diminish the reputation of the force further, thus creating a vicious cycle for military personnel planners.[5]

As seen in previous discussions, the law has not required the military to produce social science data to prove its regulatory rationale that accommodating homosexuality within the military adversely would affect recruiting, retention, and public acceptability of military service in general.[6] National opinion polls, however, do demonstrate the potential for such an adverse impact.[7] Of course, opinion polls, national surveys, and expert testimony are not the only avenues of insight into how the American public might react to the accommodation of homosexuality within its military.

Some insight is provided by the controversy surrounding homosexual clergy, homosexual marriage, and other socialization issues. These issues implicate, for example, strong state interests in ensuring that rules regarding marriage reflect the widely held values of its peoples. Thus states legitimately have established various prerequisites to marriage and banned incest, bigamy, and homosexuality.[8]

Some insight into how the American public might respond to accommodation of homosexuality in the military is provided by the scope and depth of societal disapproval of homosexual sexual practices and lifestyle, a disapproval reflected not only by the law, but also in countless instances of social tension throughout the nation. One account, for example, reported this statement from a homosexual rights event in New York: " 'I hate straights,' declared the most talked-about manifesto at New York's Gay Pride Parade on June 24. Straights, procreators, 'breeders'—that's you, heteros. Get out of our faces."[9] Protest erupted on a midwestern campus because,

> a regent . . . questioned the university's financing of a Lesbian and Gay Male Programs Office . . . [H]e requested an investigation of whether the bathrooms of a university building were "being used as a meeting place for members of the homosexual community to perform sexual acts." . . . [He] said "young people of this state, when they enroll at the university, ought not to be forced, for example, to have a professed homosexual as a roommate" . . . [The AIDS Coalition to Unleash Power saw the problem as] "institutionalized heterosexism" [and demanded] adding the term "sexual orientation" to the anti-discrimination clause of the school's bylaws, the establishment of a center for AIDS research and the creation of a lesbian-gay studies department.[10]

Sexual preference terms found in antidiscrimination clauses in bylaws and local ordinances often are flash points for public controversy. A representative example of this controversy is the "Take Back Tampa" campaign. In 1992 in Tampa, Florida, a citizens group called "Take Back Tampa" launched a referendum to repeal a local ordinance prohibiting discrimination on the basis of sexual preference. The citizens group published a campaign leaflet titled, "Sodomy is Not a Civil Right."[11] The leaflet cited four lawsuits filed by homosexual activists and aimed at forcing the state to allow homosexual couples to adopt children. The leaflet also stated Tampa homosexual activists successfully had used the sexual preference ordinance to stop the Tampa Police Department from arresting and prosecuting homosexuals who were soliciting people for sex in public parks and public restrooms.

On America's other coast, in California, similar public controversy over "special privileges" for homosexuals splits communities and taxes community resources. One commentator on Irvine, California's sexual preference ordinance states:

Sadly the [human right ordinance protecting sexual orientation]—regardless of whether the authors were well-intended or simply pandering to a narrow political interest group—has indeed been divisive and expensive. It may yet . . . be litigious and destructive.[12]

After a 10-year-old boy entered a public bathroom in a local park and found three homosexual men engaged in public sex, Irvine citizens repealed the ordinance by referendum.[13]

Other flash points across the nation center on the church. For example, a group called Artists Against Religious Oppression claimed:

The message is that the church should stay out of politics and medicine, that it should stop its oppression of women, gays and lesbians, that it should stop preaching sin and lies and that it should start preaching safe sex and the truth. . . . Witnesses said boards were bolted to the door of the Torrance church to form an approximately 10-

foot-tall cross. Affixed to the cross were about 30 penis-shaped objects . . . covered with condoms.[14]

Further insight into how the American public might respond to
the accomodation of homosexuality in the military is provided by
state and national electorates which have, with few exceptions,
rejected legislation that would have made homosexuals a protected
minority. One report indicates:

> [M]ore than 60 *towns and cities* across the country, from rural hamlets
> to metropolitan centers, . . . made it illegal to discriminate against
> gays and lesbians in such areas as employment, housing and public
> service. . . . Gay activists for the most part have been stymied in
> efforts to get anti-discrimination measures passed nationally and in
> states such as California. . . . [A]ctivists acknowledge that tracing
> [the] impact [of such ordinances] is difficult.[15]

Some argue that the lack of national legislation on behalf of
homosexuals proves homosexuals are politically powerless. As it
cannot be said homosexuals have failed "to attract the attention of
lawmakers," a more reasonable explanation for the lack of national
legislation favoring homosexuals is the simple functioning of the
democratic process—that is, the majority has declined to exercise its
prerogative to extend special protection to homosexuals.

Indeed, the Congress repeatedly has rejected attempts to amend
Title VII, the federal employment discrimination statute, to include
homosexuals. In *Ulane*, for example, a discrimination case brought
by a transsexual, the court noted that several proposed amendments to Title VII to provide coverage to homosexuals had been
rejected by Congress.[16] The court went on to find that Title VII does
not prohibit discrimination based upon "affectational or sexual orientation."[17] It is obvious "if [Title VII] permitted discrimination in
government employment that the Constitution prohibits, courts
would be obliged to hold the statute invalid to the extent it conflicted with the superior norm."[18] Title VII has not been held invalid
on the grounds it excludes homosexuals from its scope. Thus, the

national legislature's declination to afford homosexuality special legal status stands.

A common theme in the debates surrounding legislative proposals to afford special legal status to homosexuals has been the desire to avoid condoning homosexual conduct. Whether or not homosexual sexual practices are morally deserving of societal disapproval or should instead be condoned misses the point. Many disapprove of these practices, and the surrounding lifestyle. Expressions of this disapproval are found in many different forms and segments of society. One account of a recent study of attitudes toward sex and sexual mores stated:

"The purported sexual revolution of the 1960s didn't occur . . . In terms of public morality, the American population tends to be very conservative and has continued that way." . . . Arguing that America did not undergo a sexual revolution, the researchers cited data showing 72% of those interviewed said it was always wrong to have extramarital sex, *while 79% disapproved of homosexual relations.* More than 75% of those interviewed in 1970 said homosexuals should not be allowed to work as judges, schoolteachers and ministers. Almost 68% said homosexuals should not be doctors. A majority of the respondents favored laws against prostitution, homosexuality and extramarital sex.[19]

Another example is this report from the fashion world:

Garment makers don't want to offend or lose customers with a designer who flaunts his homosexuality. . . . "[L]*et's face it, [a businessman said], the majority of the public does not approve of homosexuality."* . . . Last year [1989], *74.2% of those questioned in a national survey said that sex between two adults of the same gender "is always wrong,"* according to the National Opinion Research Center in Chicago.[20]

If the wide controversy and disapproval surrounding homosexuality were projected onto the military because it accommodated homosexuals, the negative impact this outcry would have on mission accomplishment, recruiting, and retention should be obvious.

Recent reports show 61% of Americans support the military's homosexual exclusion policy[21] and the "vast majority of military members want to keep the ban."[22]

The broad view of the American people—American society—is important to the question of whether or not the homosexual exclusion policy avoids an otherwise negative impact on recruiting and retention. But the individual recruit—the young person, often a teenager, considering military service—stands at the heart of this debate.

Considering evidence that suggests nearly 75% of Americans conclude homosexuality is "always wrong,"[23]—and the observed concomitant determination to avoid sponsoring homosexuality in any way—it is naive to conclude that accommodation of homosexuality within the military would have *no* impact on public acceptability, recruiting, or retention in the armed forces. It is not mere conjecture that young people as well as their families—who are already making a serious decision on whether or not military service is right for them or their child—may be influenced to take a more cautious view of soldiering if homosexuals were—or were even perceived to be—a large, visible, or militant segment of the military community.

This hesitation—which results from doubts about how successfully homosexual conduct and the military environment can be reconciled—is not confined to the question of military service. The "cautious view" of homosexuality frequently is demonstrated in society at large. In *Rowland*, for example, one judge noted "the wrath of parents" was brought upon the school when a teacher declared she was a homosexual.[24] In another such instance, parents of 75 employees at Magic Mountain amusement park complained about a scheduled homosexual event at the theme park.[25] In New York, a whole school board suffered suspension rather than teach a curriculum that promoted "ideals [about homosexuality] opposed by parents."[26]

There is also evidence of the "social hesitation" created by the influence of homosexuality in other social institutions.[27] In 1987, the chairman of Notre Dame's theology department, for example, posed these questions in *Commonweal*, an influential theological periodical:

What impact does the presence of a large number of gay seminarians have on the spiritual tone and moral atmosphere of our seminaries? . . . How many heterosexual seminarians have decided to leave the seminary and abandon their interest in a presbyterial vocation because of the presence of significant numbers of gays in seminaries and among the local clergy? . . . Do homosexual bishops give preference, consciously or not, to gay candidates for choice pastorates?[28]

In a subsequent interview in 1990, the above author stated,

I hear about it too often from the seminary people I know, . . . [h]ow heterosexual males are being forced out, discouraged by the excessive number of homosexuals in the seminary. It was always there . . . [b]ut today the balance is being tipped in . . . favor [of homosexuals]. Claiming celibacy is a wonderful cover for gays, and let's face it, the seminary presents a marvelous arena of opportunity for them.[29]

If the results—or perceived results—of the accommodation of homosexuality in the Army followed a similar pattern—as they might—accommodation would have an adverse impact on recruiting, retention, and public acceptability of military service. Further, in terms of privacy and disciplinary risks, the soldier's lifestyle is very different from the cleric's. This difference raises the possibility that problems observed in "accommodating" homosexuals in seminaries, for example, would be magnified in the military setting.

Indeed, research on the experience of homosexuals in the Royal Dutch Navy, where homosexuals have been allowed to serve since 1974, supports the contention that recruiting would be affected by parents' reluctance "to let their teen-agers join the military if they knew the recruits could be living in close quarters with gays. . . . [One Dutch sailor said,] 'Young people in the navy are uncertain about their sexual identity.' Another [Dutch sailor] said horseplay among recruits has homosexual undertones that is disturbing to young sailors once they realize they are serving with gays."[30]

Clearly, the effect of accommodation of homosexuality in the military is not likely to be neutral as to recruiting, retention, and public acceptability. The homosexual exclusion policy takes a cautious approach, but great and unflagging caution is appropriate in dealing with the relationship between the American people and the American Army. In this regard, the principles of war are as relevant to military personnel policies as they are to battle plans—in each case the end goal is precisely the same. As Clausewitz, perhaps the most studied military strategist of all time, explained:

> The task of the military theorist . . . is to develop a theory that maintains a balance among . . . the *trinity of war*—the *people*, the *government*, and the *Army*. . . . A theory that ignores any one of them or seeks to fix an arbitrary relationship between them would conflict with reality to such an extent that for this reason alone it would be totally useless.[31]

The trinity of war—the empowering balance between the people, the government, and the Army—necessarily is a delicate balance. Accommodation of homosexuality within the military presently has the potential to disrupt this balance. Indeed, it is possible that efforts to accommodate homosexuality within the military "would conflict with [public] reality to such an extent that for that reason alone [accommodation] would be totally useless"—if not calamitous. As the law recognizes and common sense counsels, military personnel policies, like battle plans, must be infused with realism if they are to have any utility at all. The reality of the dynamic bond between the American people and the American Army—and the reality of the wide and often bitter controversy surrounding homosexuality—raises serious and real concerns about using the Army as a laboratory for experimenting with homosexual rights. As many courts have recognized, a revocation of the homosexual exclusion policy would "protect from regulation a form of behavior *never before protected*, and indeed traditionally condemned."[32]

There is a forum, however, where this controversy might properly be resolved, and where the Army may be assured that far-ranging policy decisions comport with the desires of the American people. This forum, naturally, is the national legislature. The following chapter will discuss the relationship between the coordinate branches of government and the role and power of the legislature to "raise and support the land and naval forces."[33]

CHAPTER EIGHT

The Relationship Between the Coordinate Branches of Government

CONSTITUTIONAL REASONS FOR POLICY-MAKING RESTRAINT

Judges are not given the task of running the Army.
—Orloff v. Willoughby

"Judges are not given the task of running the Army."[1] This succinct explication of constitutional order was made by the Supreme Court in 1953. Before and since, the Court consistently and forcefully has emphasized that judicial deference to professional military judgments is an unwavering requirement for proper functioning of the three coordinate branches of government.[2]

The requirement that judges recognize and defer to military judgment and expertise is based partially on the recognition that "it is difficult to conceive of an area of governmental activity in which the courts have less competence [than the] complex, subtle, and professional decisions as to the composition, training, equipping and control of a military force."[3] This practical basis for judicial deference proves itself. More critical than the practical basis for judicial deference, however, is its constitutional basis, which is founded on separation of governmental powers among the coordinate branches.

Judicial deference toward military judgments is constitutionally required because "ultimate responsibility for [the military] is appropriately vested in branches of the government which are periodically subject to electoral accountability."[4] The potential result of departing from this standard is severe. As the Supreme Court strongly has counseled, it is this very "power of oversight and control of the military force by elected representatives and officials *which underlies our entire constitutional system*."[5] Harry Summers, a noted military analyst, observes, *"The Federalist Papers* clearly show . . . the Founding Fathers deliberately rejected the idea of an 18th century-type Army *answerable only to the Executive.* They wrote into the Constitution *specific safeguards to ensure the people's control of the military*."[6]

When these constitutional principles of separation of powers are abandoned or manipulated, the result is a military unusually vulnerable to the agendas—even whims—of those who view the Army as an engine for social change, a leader in social experimentation and adaptability to changing community standards, the revisionist of traditional wisdom and values, a benevolent welfare agency, or a college for career or professional development. In such hands, the relationship between the coordinate branches of government breaks down and the relationship between the nation and her Army breaks down.

Opponents of the homosexual exclusion policy have challenged the judiciary not only to overstep its bounds, but also to abandon the constitutional principle of the military's accountability to the elected Congress. Plaintiffs have urged judges to substitute their judgment for the judgment of Congress and the military's senior leadership, who are accountable to Congress. The *Ben-Shalom* court recognized this challenge and rejected it, explaining:

[T]he Army should not be required by this court to assume the risk, *a risk it would be assuming for all our citizens*, that accepting admitted homosexuals into the armed forces might imperil morale, discipline, and the effectiveness of our fighting forces. The Commander-in-

Chief, the Secretary of Defense, the Secretary of the Army, and the generals have made the determination about homosexuality . . . and we, as judges, should not undertake to second-guess those with the direct responsibility for our armed forces.[7]

The scope and complexity of the issues presented by accommodating homosexuality in the Army—rather than excluding homosexuals from military service—demonstrate the wisdom of judicial restraint and judicial deference to the military. Conversely, these issues reinforce the constitutional and practical imperative that the democratic process be brought to bear in making risky policy decisions with far-ranging implications for force composition.

The value and utility of the democratic process to resolve the controversial and far-ranging issues raised by challenges to the homosexual exclusion policy is often overlooked. Plaintiffs challenged the courts to change the policy from the bench. Some commentators have challenged the service secretaries to change the policy from the Pentagon. Some have even called for the President to rescind the policy by Executive Order, and President Clinton has indicated a willingness to exercise this option. Those who counsel judges or military officials or even the President to act with dispatch to rescind the homosexual exclusion policy should recall this fact: "[t]he American Army really is a people's Army in the sense that it belongs to the American people, who take a jealous and proprietary interest in its involvement."[8] The people exercise their control—their jealous and proprietary interest—through the elected Congress, and it is there that the power of decision in bitterly controversial force composition issues properly lies.[9]

The Congress has addressed the issue of homosexuals in federal employment. Soldiering, of course, is not "federal employment" in the strict sense, and soldiers are not in an employee-employer relationship with the Army. Rather, recruitment into the military effects a change in status from civilian to soldier.[10] Nevertheless, Congress's actions in regard to homosexuality within the civil service are instructive.

For example, Congress repeatedly has rejected proposals to

amend Title VII, the federal employment discrimination statute, to extend coverage to homosexuals.[11] If subordinate, or even executive, policy-makers rescinded the homosexual exclusion policy on their own initiative, they would be providing legislative status to homosexuality within the military while the Congress has not seen fit to provide such status to homosexuality within the federal civilian service or elsewhere.[12] Such policy initiatives would "protect from regulation a form of behavior never before protected, and indeed traditionally condemned."[13]

That result would not sensibly account for the differences between soldiering and federal service, much less for the concerns expressed in the regulatory rationales for the homosexual exclusion policy—concerns which are unique to the military environment. It might instead "fix an arbitrary relationship between [the people, the government, and the Army that] conflict[s] with reality"—that is, the reality of the present legal and social status of homosexuality in America.

This "reality" often has been a creature of personal surmisal rather than fact. The *Matlovich* court, for example, in 1978 charged that the Army should be the "leader in social experimentation and . . . adaptability to changing community standards."[14] The court assumed—in 1978, as often is assumed today—*that community standards on homosexuality were in fact "changing."* There exists evidence to the contrary.[15] Moreover, even if community standards—at some level of social organization—were changing in 1978, there is no evidence in the 1990's that community standards have changed so radically that homosexuality is now an accepted standard of conduct in American society and institutions, including the law. Thus, it would be arbitrary to fix a special relationship between the homosexual community and the military that had no support in the law and no counterpart in other social institutions.

The court in *Dronenburg* simply said, "[i]f the revolution in sexual mores that appellant proclaims is in fact ever to arrive, we think it must arrive through the *moral choices of the people and their elected representatives,* not through . . . this court."[16] This is excellent constitutional and practical advice. To seize the initiative away from the Congress and make the military a proving ground for homosexual social and political issues—particularly when the larger society is nowhere near a consensus in favor of such issues—would be a disturbing anomaly, and that result should be avoided.

Myriad, complex, and emotional issues attend proposals to accommodate homosexuality, whether in the military or in any other setting. Only the democratic process provides the appropriate forum for controversy of this scope. Only the democratic process can gauge truly the pulse of the public and, thus, best judge the competing interests of the individual and a combat-ready Army.

As Justice Scalia keenly has observed, "[o]ur salvation is the Equal Protection Clause, which requires the democratic majority to accept for themselves and their loved ones what they impose on you and me."[17] Clearly, reliance on the democratic process is the virtue and strength of the rational basis test under the Equal Protection Clause. Applied to the homosexual exclusion policy, the democratic process would ensure that the nation is willing to sign up for and support the type of army that policy-makers create by military personnel decisions with the potential for far-ranging effects on force composition—indeed, with the potential for far-ranging effects on society and our nation as a whole.

There is ample evidence, in both law and fact, as well as in common sense and experience, to support the secretarial determination that homosexuality is incompatible with military service. To make a determination to the contrary is fraught with risk. Permitting homosexuals to serve—and accommodating homosexuality within the military—poses risks to good order, discipline, and morale, risks to security and privacy concerns, and risks to the public acceptability and desirability of military service. It also risks dragging the Army onto heated social battlefields and hailing it into court constantly to defend its policy choices.

In short, a determination contrary to the homosexual exclusion policy risks the combat readiness of the armed forces. Such weighty risk should be assumed only after the most thoughtful and thorough consultation with the American people. After all, it is their army. And it is their prerogative, through their elected representatives, to "not only raise, support, and regulate the Army but also [to] determine the very battlefields upon which their Army must fight."[18]

CHAPTER NINE ·

———

Summary

As General MacArthur observed in his famous address to the cadets at West Point, which has become an address to every soldier who has ever served,

> through all this welter of [social] change and development, your mission remains fixed, determined, inviolable—*it is to win our wars.* Everything else in your professional career is but corollary to this vital dedication. All other public purposes, all other public projects, all other public needs, great or small, will find others for their accomplishment; *but you are the ones who are trained to fight; yours is the profession of arms.*[1]

The profession of arms is unlike any other. It is not a profession of individual aspiration, nor is it a profession of social pro-action. In evaluating military personnel policies or decisions about the composition of the armed forces, the touchstone cannot be how the policy or decision furthers the interest of the individual or the enlightenment of society at large.

From the Army's point of view, the proper first question in such decisions is not how the decision will affect single parents, handicapped persons, homosexuals, women, minorities, and so on, but rather: "Will this policy decision *make us a better army?* Will it im-

187

prove the Army's ability to move, shoot, and communicate? Or, put simply, will it enhance combat readiness?"

If the answer to that question is a resounding "yes," then the policy should be implemented as soon as possible. If the answer to that question is "no" or only "perhaps not," then grave trepidation is in order.

A substantial and virtually unanimous body of law affirms the constitutionality of the military's homosexual exclusion policy. A substantial body of medical and other literature demonstrates a factual basis for the secretarial determination that homosexuality is incompatible with military service.

Allowing homosexuals to serve in the military may somehow benefit individual homosexuals. It may benefit the homosexual "cause." Indeed, because of the unique nature of military service, every other question of homosexual employment rights may be cast as a lesser included question in the issue of whether or not homosexuality is compatible with military service. There is no ground whatsoever upon which to conclude, however, that allowing homosexuality in the military would "make us a better army."

Homosexuals themselves repeatedly have conceded the detrimental and disruptive effect of homosexual conduct within the military.[2] Even when lower courts struck down the homosexual exclusion policy (decisions later reversed in every case), there were careful remarks that "the Court does not hold that the [military] is constitutionally required to enlist or retain persons who engage in homosexual acts."[3]

This very realistic concession—that it would be proper to exclude homosexuals who, unlike plaintiffs, committed homosexuals acts— demonstrates the absolute criticality of applying an *ex ante* analysis to policy-making.[4] This easily is the most important point of this whole discussion. The Army cannot tell any better than the individual whether he will give in to temptation or obey regulations. Moreover, search though one might, there is no evidence that homosexual conduct is a remote or infrequent possibility within the homosexual community.

The question, "would a policy decision allowing homosexuals to serve make us a better army?", is an *ex ante* question. An army where some homosexuals may serve covertly by evading the exclusion policy is not the same as an army where all homosexuals are allowed to serve openly. An army where some soldiers break the rules is not the same as an army where the rules are removed. The Army must count the cost of a policy decision in terms of combat readiness.

Experimenting with military personnel management and military force composition costs. It costs money, personnel resources, and time. It costs administrative burdens and the potential for expensive and protracted litigation. It costs—at least initially—some degree of diminishment in good order, discipline, and morale, hence in combat readiness. In regard to experimenting with the homosexual exclusion policy, these costs to the Army simply are unwarranted and wasteful. Similarly to other groups that are excluded from military service, no evidence exists that the positive contribution of individual homosexuals would outweigh the potential of homosexuals as a group and homosexuality as a sub-culture to detract from successful military service or from the good order, discipline, and morale of the fighting force as a whole.

Resources, tangible and intangible, are too precious to gamble on military personnel policies. And this nation's Army, which history shows to be at once strong and fragile, plainly is too precious to gamble on social experiments and social initiatives that are far beyond those presently embraced by the nation this Army serves. For now, at least, the homosexual exclusion policy strikes an appropriate balance between the desires of individuals and the requirements of "this nation's war-guardian."

The United States no doubt will continue to grapple with issues of homosexual rights. But to the extent homosexuals advance their cause as one of the nation's "public purposes . . . public projects . . . or public needs," it is appropriate for homosexuals to "find others [besides the military] for their accomplishment." There is no sensible or sufficient reason for the military—as a social and constitutional institution, much less as a warfighting force—to attempt to go first in forging rights for homosexuals. Social reformation is neither the province nor the mission of America's fighting force. Rather, the mission of America's fighting force is a constant, costly, and consuming vigilance—both internal and external—that is unknown outside the profession of arms. Indeed, as General MacArthur poignantly observed, the military has a single, fixed, determined, inviolable mission—*"it is to win our wars."*

Notes

Chapter One

1. Crocker, The Army Officer's Guide, ix (1981).
2. This book addresses the Army regulation on homosexuality, which is based on Department of Defense policy. The book is written with references to the Army and soldiers. The services, however, have similar homosexual exclusion policies. The analysis set out in this book may be equally applicable to the other military branches.
3. *E.g.*, Department of Defense Directive 1332.14, Enlisted Administrative Separations (Jan. 28, 1982). Regulatory references are to Army Regulation [AR] 635-200, Enlisted Ranks Personnel, chap. 15 (1990). Other pertinent regulations are cited *infra* chap. 5, note 7.
4. *See, e.g.*, Lindenau v. Alexander, 663 F.2d 68 (10th Cir. 1981); *cf.* Mack v. Rumsfeld, 609 F. Supp. 1561 (W.D.N.Y. 1985).
5. Alberico v. United States, 783 F.2d 1024 (Fed. Cir. 1986); *see* United States v. Fisher, 477 F.2d 300 (4th Cir. 1973) (defendant could not enlist voluntarily in the military because he was pending a felony charge).
6. Smith v. Christian, 763 F.2d 1322 (11th Cir. 1985) (applicant excluded from military service because he was missing a finger).
7. Doe v. Alexander, 510 F. Supp. 900 (D. Minn. 1981).
8. Johnson v. Robison, 415 U.S. 361 (1974) (conscientious objectors were excluded from military service; they later sued because they were denied veterans' benefits).
9. *See, e.g.*, Doe v. Garrett, 903 F.2d 1455 (11th Cir. 1990) (upholding military exclusion for human immunodeficiency virus [HIV] sero-positivity). Authority to set medical standards for military accession and retention is at 10 U.S.C. sec. 505(a).

10. Williams v. United States, 541 F. Supp. 1187 (E.D.N.C. 1982)(Marine discharged for being 50 pounds over military weight standards); Vance v. United States, 434 F. Supp. 826 (N.D. Tex. 1977); cf. Vanguard Justice Soc'y v. Hughes, 471 F. Supp. 670 (D. Md. 1979).
11. See Billings v. Truesdell, 321 U.S. 542 (1944) (citing physical and mental aptitude requirements under selective service legislation).
12. See United States v. Lavin, 346 F. Supp. 76 (S.D.N.Y. 1972).
13. Blameuser v. Andrews, 630 F.2d 538 (7th Cir. 1980) (Reserve Officer Training Corps (ROTC) cadet excluded from ROTC for adhering to nazism); Khalsa v. Weinberger, 759 F.2d 1411 (9th Cir. 1985) (member of Sikh religion excluded from the military because his religious precepts precluded him from complying with military appearance regulations); cf. Goldwasser v. Brown, 417 F.2d 1169 (D.C. Cir. 1969) (discharge of a civilian employee of the Air Force for discussing personal political views on the Viet Nam war was permissible).
14. See, e.g., Dicicco v. Immigration and Naturalization Serv., 873 F.2d 910 (6th Cir. 1989) (plaintiff was originally excluded from military service because he did not speak English).
15. See, e.g., Spain v. Ball, 1991 U.S. App. LEXIS 4127 (2d Cir. 1991) (military age requirements imposed by 10 U.S.C. sec. 532 (1988) are permissible).
16. E.g., Alberico, 783 F.2d 1024; Crawford v. Cushman, 531 F.2d 1114 (2d Cir. 1976); Lindenau, 663 F.2d at 72; Pauls v. Sec'y of Air Force, 457 F.2d 294 (1st Cir. 1972); Doe v. Alexander, 510 F. Supp. 900.
17. See, e.g., Kennedy v. Mendoza-Martinez, 372 U.S. 144 (1942).

Chapter Two

1. The text refers to reported cases that reached substantive challenges to the homosexual exclusion policy. The cases were Ben-Shalom v. Marsh, 881 F.2d 454 (7th Cir. 1989), cert. denied, 110 S. Ct. 1296 (1990); Watkins v. United States Army, 847 F.2d 1329 (9th Cir. 1988), vacated on other grounds, 875 F.2d 699 (9th Cir. 1989) (en banc); Matthews v. Marsh, 755 F.2d 182 (1st Cir. 1985); Rich v. Secretary of the Army, 735 F.2d 1220 (10th Cir. 1984); Pruitt v. Weinberger, 659 F. Supp. 625 (C.D. Cal. 1987); cf. Krugler v. United States Army, 594 F. Supp. 565 (N.D. Ill. 1984); Von Hoffburg v. Alexander, 615 F.2d 633 (5th Cir. 1980) (dismissed for failure to exhaust administrative remedies).
2. The text refers to reported cases that reached substantive challenges to

the homosexual exclusion policy. Woodward v. United States, 871 F.2d 1068 (Fed. Cir. 1989); Dronenburg v. Zech, 741 F.2d 1388 (D.C. Cir. 1984); Beller v. Middendorf, 632 F.2d 788 (9th Cir. 1980); Berg v. Claytor, 591 F.2d 849 (D.C. Cir. 1978); Steffan v. Cheney, 733 F. Supp. 121 (D.D.C. 1989); Johnson v. Orr, 617 F. Supp. 170 (E.D. Cal. 1985); *cf.* Matlovich v. Secretary of the Air Force, 591 F.2d 852 (D.C. Cir. 1978); Secora v. Fox, 747 F. Supp. 406 (S.D. Ohio 1989); Doe v. Secretary of the Air Force, 563 F. Supp. 4 (D.D.C. 1982) (decided on procedural due process).

3. M. Humphrey, My Country, My Right to Serve, 94, 108, 122, 139, 157, 167-68, 170, 243-44 (1990).

4. *Id.* at 108 (interview with Hatheway). *See* Hatheway v. Sec'y of the Army, 641 F.2d 1376 (9th Cir. 1981).

5. *Id.* at *xvii* & *xviii.*

6. *Id.* at 243-44.

7. Dubbs v. Central Intelligence Agency, 866 F.2d 1114 (9th Cir. 1989).

8. Dillard v. Brown, 652 F.2d 316 (3d Cir. 1981).

9. *But see Woodward,* 871 F.2d 1068.

10. *Watkins,* 875 F.2d 699; *cf.* M. Humphrey, *supra* chap. 2, note 3, at 34 & 73.

11. *Watkins,* 875 F.2d 699.

12. M. Humphrey, *supra* chap. 2, note 3, at 73 (interview with Berg) [original emphasis deleted and emphasis added]; *see Berg,* 591 F.2d 849.

13. United States v. Matthews, 38 C.M.R. 430 (C.M.A. 1968).

14. *But see Watkins,* 847 F.2d 1329, *vacated on other grounds,* 875 F.2d 699 (9th Cir. 1989) (en banc).

15. *See, e.g., Lindenau,* 663 F.2d at 73 (rejecting claim single parent exclusion policy infringed plaintiff's privacy in matters of marriage or child bearing).

16. Bowen v. Gilliard, 483 U.S. 587, 601-02 (1987) (rejecting claim exclusion from statutory entitlements scheme infringed plaintiff's privacy in family matters).

17. *Ben-Shalom,* 489 F. Supp. 964, 975 (E.D. Wis. 1980) (*Ben-Shalom I*).

18. *Dronenburg,* 741 F.2d at 1392.

19. *See* Bowers v. Hardwick, 478 U.S. 186, 190 (1986).

20. *Matthews,* No. 82-0216-P, slip op. (D. Me. Apr. 3, 1984).

21. Prince v. Massachusetts, 321 U.S. 158 (1944).

22. Loving v. Virginia, 388 U.S. 1 (1967).

23. *Griswold,* 381 U.S. 479; Eisenstadt v. Baird, 405 U.S. 438 (1972) (contraception); Roe v. Wade, 410 U.S. 113 (1973); Carey v. Population Servs.

Int'l, 431 U.S. 678 (1977) (abortion); Skinner v. Oklahoma ex rel. Williamson, 316 U.S. 535 (1942) (procreation); *cf. Lindenau*, 663 F.2d 68; West v. Brown, 558 F.2d 757 (5th Cir. 1977) (military policy of excluding single parents from service did not curtail freedom to choose in matters of marriage, family life, or child bearing); *see also Dillard*, 652 F.2d 316.

24. *E.g., Hardwick*, 478 U.S. at 194-95.
25. *Id.; see Woodward*, 871 F.2d 1068.
26. Bradbury v. Wainwright, 718 F.2d 1538, 1540 (11th Cir. 1983) (citing Sosna v. Iowa, 419 U.S. 393, 404 (1975); Zablocki v. Redhail, 434 U.S. 374, 399 (1978) (Powell, J., concurring)); *cf.* United States v. Wheeler, 30 C.M.R. 387 (C.M.A. 1961) (upholding military regulations on marriage).
27. *E.g., Rich*, 735 F.2d 1220; *Dronenburg*, 741 F.2d 1388.
28. *See* High Tech Gays v. Defense Indus. Security Clearance Office, 909 F.2d 375 (1989) (Canby, J., dissenting from rejection of suggestion of rehearing en banc) (suggesting the Equal Protection Clause is to protect people from discrimination based on what they are); *accord* Davis, *Military Policy Toward Homosexuals: Scientific, Historical, and Legal Perspectives*, 131 Mil. L. Rev. 55, 91 (1991).
29. *Compare Ben-Shalom I*, 489 F. Supp. 964; *Matthews*, No. 82-0216-P, slip op. (D. Me. Apr. 3, 1984); Baker v. Wade, 553 F. Supp. 1121 (N.D. Tex. 1982) *with Rich*, 735 F.2d 1220; *Dronenburg*, 741 F.2d 1388; Baker v. Wade, 769 F.2d 289 (5th Cir. 1985); *see Hardwick*, 478 U.S. 186 (lower courts often accepted broad privacy arguments, but appellate courts narrowly construed the right to privacy).
30. *Rich*, 735 F.2d at 1228; *see also Hatheway*, 641 F.2d at 1383-84 (personal autonomy not violated when soldier was given a dishonorable discharge for homosexual acts); *Dronenburg*, 741 F.2d 1388; *Matlovich*, 591 F.2d 852; *see also Matthews*, 755 F.2d 182; *Baker*, 769 F.2d 289; *but see* New York v. Onofre, 415 N.E.2d 936 (1980) (holding criminal proscription of sodomy unconstitutional).
31. *See, e.g.,* Jacobson v. Massachusetts, 197 U.S. 11 (1905) (vaccination).
32. *Hardwick*, 478 U.S. at 190.
33. *Ben-Shalom I*, 489 F. Supp. at 975.
34. *Rich*, 735 F.2d at 1228.
35. *Ben-Shalom I*, 489 F. Supp. at 975.
36. *Matthews*, No. 82-0216-P, slip op. (D. Me. Apr. 3, 1984) (viewing homosexual status as distinct from homosexual propensity).

37. *Ben-Shalom*, 881 F.2d at 456.
38. United States v. Voorhees, 16 C.M.R. 83, 106 (C.M.A. 1954) (Latimer, J., concurring in part and dissenting in part).
39. *See, e.g., Ben-Shalom*, 881 F.2d at 454.
40. *Lindenau*, 663 F.2d at 73; *West*, 558 F.2d 757; *Dillard*, 652 F.2d 315; *cf. Gilliard*, 483 U.S. 587, 601-02.
41. Dallas v. Stanglin, 490 U.S. 19, 23-24 (1989).«LB«PG»»
42. *Id*. at 24 (quoting Roberts v. United States Jaycees, 468 U.S. 609, 617-18 (1984).
43. *Id*. at 25; *cf*. Board of Directors of Rotary Int'l v. Rotary Club of Duarte, 481 U.S. 537, 548 (1987) (association for the purpose of public debate and similar protected purposes).
44. *Id*. at 24, 15; Stanley v. Georgia, 394 U.S. 557 (1969).
45. *Berg*, 591 F.2d at 849.
46. *Dronenburg*, 741 F.2d at 1388.
47. Plaintiffs stated they were homosexual and numerous homosexual acts were established on the record. *Berg*, 591 F.2d at 850; *Dronenburg*, 741 F.2d at 1398; *see also* M. Humphrey, *supra* chap. 2, note 3, at 72-79 (interview with Berg); *id*. at 89-92 (interview with Dronenburg).
48. *Berg*, 436 F. Supp. at 79.
49. *Ben-Shalom*, 881 F.2d at 464.
50. *Woodward*, 871 F.2d at 1074, n.6.
51. *Pruitt*, 659 F. Supp. at 626.
52. *Steffan*, 733 F. Supp. at 121; *see Matthews*, 755 F.2d at 183.
53. Aumiller v. University of Del., 434 F. Supp. 1273, 1303 & n.86 (D. Del. 1977).
54. *See, e.g.*, M. Humphrey, *supra* chap. 2, note 3, at 235-43, 161-67.
55. Cyr v. Walls, 439 F. Supp. 697, 702 (N.D. Tex. 1977).
56. *Ben-Shalom I*, 489 F. Supp. at 964.
57. *Matthews*, No. 82-0216-P, slip op. (D. Me. Apr. 3, 1984).
58. 871 F.2d 1068.
59. Woodward subsequently admitted in interviews to homosexual acts. M. Humphrey, *supra* chap. 2, note 3, at 161-67.
60. *Ben-Shalom I*, 489 F. Supp. at 974.
61. Healy v. James, 408 U.S. 169, 192 (1972).
62. *Id*. at 971. National Gay Task Force v. Oklahoma, 759 F.2d 1270 (10th Cir. 1984); *see Cyr*, 439 F. Supp. at 700 (homosexuals "have the fundamental right to meet, discuss current problems, and to advocate changes in the status quo, so long as there is no 'incitement to immi-

nent lawless action' "); *cf.* Weston v. Lockheed Missiles and Space Co., 881 F.2d 814 (9th Cir. 1989) (plaintiff stated he was a member of an organization called Lesbian and Gay Associated Engineers and Scientists; Lockheed did not submit plaintiff's security clearance application because Lockheed concluded the application revealed "evidence of homosexuality").

63. Davis, *supra* chap. 2, note 28, at 55.

64. *Ben-Shalom*, 881 F.2d at 460 (footnotes omitted); *see Watkins*, 875 F.2d at 707.

65. *Id.* at 460; *see Dronenburg*, 746 F.2d 1579, 1582, n.1 (suggestion for rehearing en banc denied) (Bork, Scalia, J.J.).

66. United States v. Harriss, 347 U.S. 612, 626 (1954); *see* United States v. O'Brien, 391 U.S. 367 (1968).

67. *E.g.*, *Ben-Shalom*, 881 F.2d at 458-60 (the homosexual exclusion policy clearly promotes a legitimate government interest).

68. This section will address free speech issues. In addition to challenges based on theories of free speech, privacy and association, First Amendment challenges were based on the right to free exercise of religion and the Establishment Clause. *See, e.g., Hatheway*, 641 F.2d at 1378 ("introduced affidavits to the effect that sodomy prohibitions have religious origins and that homosexual acts, standing alone, are not harmful"). These claims failed. *Id.*

69. *O'Brien*, 391 U.S. at 375 [emphasis added].

70. Connick v. Myers, 461 U.S. 138, 143 (1983).

71. *Id.*

72. *Id.* at 377; *see* Arcara v. Books, 478 U.S. 697 (1986).

73. *See, e.g., Johnson*, 617 F. Supp. at 173; *Pruitt*, 659 F. Supp. at 627; *Ben-Shalom*, 881 F.2d at 462; *see also Woodward*, 871 F.2d at 1074, n.6; *Rich*, 735 F.2d at 1224, 1225.

74. *Ben-Shalom*, 881 F.2d at 462; *accord Johnson*, 617 F. Supp. at 170.

75. *Johnson*, 617 F. Supp. at 172-73.

76. *O'Brien*, 391 U.S. at 376.

77. *See, e.g.*, AR 635-200, chap. 5 (separation from service for failure to meet height and weight standards).

78. *Blameuser*, 630 F.2d 538; *cf. O'Brien*, 391 U.S. at 370 (conviction for burning draft card as an expression of political beliefs did not abridge the First Amendment); *Goldwasser*, 417 F.2d 1169 (D.C. Cir. 1969), *cert. denied*, 397 U.S. 922 (1970) (upholding Air Force discharge of a civilian

employee for discussing "religion, politics, [and] race" while teaching English to foreign military officers).

79. *Pruitt*, 659 F. Supp. at 627; *cf. Khalsa*, 759 F.2d 1411 (upholding Army's refusal to process enlistment application of individual who stated he was a Sikh, because applicant's religious beliefs made complying with Army grooming and appearance regulations impossible).

80. *Johnson*, 617 F. Supp. at 172.

81. *Connick*, 461 U.S. 138. Soldiers, though employed by the government, have a unique status that permits more stringent regulation of speech than permitted for non-military government employees.

82. *Id.* at 145-47; *accord*, Pickering v. Board of Educ., 391 U.S. 563 (1968); Roth v. United States, 354 U.S. 476 (1956).

83. *E.g.*, *Johnson*, 617 F. Supp. at 178; *Pruitt*, 659 F. Supp. at 627; *Ben-Shalom*, 881 F.2d at 462.

84. *Connick*, 461 U.S. at 147-48.

85. *E.g.*, *Woodward*, 871 F.2d at 1071, n.2.

86. *Matthews*, No. 82-0216-P, slip op. (D.C. Me. Apr. 3, 1984), *remanded* 755 F.2d 182.

87. *See Rowland*, 470 U.S. at 1012 (Brennan and Marshall, J.J., dissenting).

88. *Roth*, 354 U.S. at 484 [emphasis added].

89. *O'Brien*, 391 U.S. at 376.

90. *Rich*, 735 F.2d at 1229, citing Brown v. Glines, 444 U.S. 348, 354 (1980) (quoting Schlesinger v. Councilman, 420 U.S. 738, 757 (1975)).

91. One plaintiff unsuccessfully challenged the proscription of sodomy as applied to him. *Hatheway*, 641 F.2d at 1381; *see, e.g.*, United States v. Scoby, 5 M.J. 160 (C.M.A. 1978); United States v. Lovejoy, 42 C.M.R. 210 (C.M.A. 1970) (military law proscribing sodomy is constitutional).

92. *See, e.g.*, *Matthews*, 755 F.2d at 183 (plaintiff did not challenge Army's right to disenroll her from ROTC for homosexual acts); *Ben-Shalom*, 881 F.2d at 461 ("[i]t is not disputed, the plaintiff and the district court agree, that the Army has the right to enforce its regulation prohibiting the actual performance of homosexual actions").

93. *Steffan*, 733 F. Supp. at 123 [emphasis added].

94. *Steffan*, 420 F.2d 74 (D.C. Cir. 1990).

95. *See* M. Humphrey, *supra* chap. 2, note 3, at 235-43.

96. *Ben-Shalom*, 881 F.2d at 459-61.

97. *See, e.g.*, *Rowland*, 730 F.2d 444, n.13 ("jury is entitled to make rational inferences and apply its common-sense knowledge of the world").

98. Author's notes from oral argument (May 18, 1989) (available from the U.S. Army Litigation Division, Ballston Metro Building, Arlington, VA).

99. AR 635-200, para. 15-3a(note).

100. *Ben-Shalom*, 881 F.2d at 457; *cf. Steffan*, 733 F. Supp. 121 (plaintiff did not challenge Navy's right to refuse reinstatement on grounds of commission of homosexual acts, but invoked Fifth Amendment to avoid answering discovery questions about homosexual conduct).

101. Robinson v. California, 370 U.S. 660, 664 (1962).

102. *Id.* at 665.

103. *Id.* at 666.

104. *Id.* at 667, n.9.

105. Mathews v. Eldridge, 424 U.S. 319 (1976) (the amount of process that is "due" depends on the consequences of the deprivation).

106. Cruzan v. Missouri, 110 S. Ct. 2841 (quoting Santosky v. Kramer, 455 U.S. 745, 756 (1982)) [emphasis added].

107. Addington v. Texas, 441 U.S. 418, 423 (1979) (Harlan, J., concurring).

108. *Cf. Gilliard*, 483 U.S. 587 (exclusion from welfare benefits scheme was not punitive).

109. *Rich*, 735 F.2d at 1224-25, n.1; *Pruitt*, 659 F. Supp. at 627.

110. *Robinson*, 370 U.S. at 664, 666 [emphasis added].

111. *Id.* at 667, n.9.

112. *See Rich*, 735 F.2d at 1228.

113. *Cf. Rich*, 735 F.2d at 1228 (even if privacy rights were implicated by the homosexual exclusion policy, that did not render the policy unconstitutional).

114. *Ben-Shalom*, 881 F.2d at 464.

115. AR 635-200, para. 15-2a.

116. M. Humphrey, *supra* chap. 2, note 3.

117. *Padula*, 822 F.2d at 102.

118. *High Tech Gays*, 909 F.2d at 380 (Canby, J., dissenting); *see also Woodward*, 871 F.2d at 1076, n.10.

119. *Baker*, 553 F. Supp. 1121 (citing Head v. Newton, 596 S.W.2d 209 (Tex. Civ. App. 1980).

120. Surawicz, *Intestinal Spirochetosis in Homosexual Men*, 82 Am. J. Med. 587-92 (1987).

121. The American Heritage Dictionary of the English Language (ed. Morris) American Heritage Publishing Co., Inc. and Houghton Mifflin Company, Boston 1975.

122. *See, e.g., Ben-Shalom,* 881 F.2d at 460; *Woodward,* 871 F.2d at 1069-70; *Dronenburg,* 741 F.2d at 1389; *Pruitt,* 659 F. Supp. at 626; *Johnson,* 617 F. Supp. at 171.

123. *Ben-Shalom,* 881 F.2d at 464.

124. Harper v. Wallingford, 877 F.2d 728, 730 (9th Cir. 1989) [emphasis added].

125. *See Ben-Shalom,* 881 F.2d at 464.

126. *Pruitt,* 659 F. Supp. at 625; *Rich,* 735 F.2d at 1224 (declamations of homosexuality have the same adverse implications for good order, discipline, and morale as do homosexual acts).

127. *Ben-Shalom,* 881 F.2d at 462.

128. *Id.; cf. Harriss,* 347 U.S. at 626 (even if the statute restrained speech, the restraint was the indirect result of self-censorship).

129. *Id.* at 460-61 ("the Army does not have to take the risk that an admitted homosexual will not commit homosexual acts which may be detrimental to its assigned mission"); *accord Pruitt,* 659 F. Supp. 625; *cf. Harper,* 877 F.2d 728; Espinoza v. Wilson, 814 F.2d 1093 (6th Cir. 1987) (prison officials did not have to assume the risk that homosexually oriented literature would have an adverse impact on discipline and rehabilitative efforts within the prison); *see also Dubbs,* 866 F.2d at 1118 (CIA Director testified homosexuality raises a risk that must be resolved in favor of the agency); *Gilliard,* 483 U.S. at 603, n.9 (recognizing the statutory risk allocation between the State and the individual).

130. *O'Brien,* 391 U.S. 367 (1968).

131. *Id.* at 377.

132. *E.g., Rich,* 735 F.2d at 1224-25, n.1 (the Secretary has the authority to set military accession standards such as the homosexual exclusion policy).

133. *See, e.g., id.* at 1227-28, n.7 (the Army's regulatory justifications are sufficient to sustain the homosexual exclusion policy); *Woodward,* 871 F.2d at 1076 (the homosexual exclusion policy serves legitimate state interests of good order, discipline, and morale within the armed forces); *accord Beller,* 632 F.2d 788.; *cf.* Metro Broadcasting, Inc. v. Federal Communications Comm'n, 110 S. Ct. 2997 (1990) (minority ownership policies, which effectively may exclude some non-minority owners, serve an important governmental interest in dissemination of diverse information to the public).

134. *Glines,* 444 U.S. at 358.

135. *Goldman*, 475 U.S. 503.
136. *O'Brien*, 391 U.S. at 377.
137. *See, e.g.*, Parker v. Levy, 417 U.S. 733 (1974).
138. United States v. Priest, 45 C.M.R. 338, 344 (C.M.A. 1972).
139. *Voorhees*, 16 C.M.R. at 106.
140. Petrey v. Flaugher, 505 F. Supp. 1087, 1091 (E.D. Ken. 1981) (student challenged expulsion from school for smoking marijuana).
141. *Johnson*, 617 F. Supp. at 175.
142. United States v. Chad, 82 C.M.R. 483 (C.M.A. 1963).
143. United States v. Vanderwier, 25 M.J. 263 (C.M.A. 1987).
144. *Ben-Shalom*, 881 F.2d at 461 (quoting *Goldman*, 475 U.S. at 507 [emphasis added]).
145. *See Dronenburg*, 741 F.2d at 1397, n.6 (discussing bestiality as a sexual preference); *see infra* chap. 4, notes 140 to 143 and accompanying text.

Chapter Three

1. *Hardwick*, 478 U.S. at 191.
2. *Id.* at 194.
3. *Cruzan*, 110 S. Ct. 2841 (1990) (Scalia, J., concurring).
4. Poe v. Ullman, 367 U.S. 497, 543 (1961) (Harlan, J., dissenting).
5. *Hardwick*, 478 U.S. at 192; *see, e.g.*, *Cruzan*, 110 S. Ct. 2841 (Scalia, J., concurring); Moore v. East Cleveland, 431 U.S. 494, 502-03 (1977) (plurality opinion).
6. *Hardwick*, 478 U.S. at 192, 197 (Burger, C.J., concurring).
7. *Cruzan*, 110 S. Ct. 2841 (Scalia, J., concurring) [emphasis added].
8. *Lindenau*, 663 F.2d at 72.
9. *Beller*, 632 F.2d at 807.
10. *Id.* at 810; *but see High Tech Gays*, 895 F.2d at 572 (the heightened solicitude test was overruled by *Hardwick*).
11. Norton v. Macy, 417 F.2d 1161, 1167 (D.C. Cir. 1969).
12. Miller v. Rumsfeld, 647 F.2d 80 (9th Cir. 1981) (Norris, J., dissenting from rejection of suggestion for rehearing en banc).
13. *Id.* at 85 [emphasis added].
14. *See Ben-Shalom*, 881 F.2d at 465.
15. *Hardwick*, 478 U.S. at 192.
16. *Baker*, 769 F.2d at 292 (quoting Berman v. Parker, 348 U.S. 26, 32 (1954)).
17. M. Humphrey, *supra* chap. 2, note 3, at *xxii*.

18. Davis, *supra* chap. 2, note 28, at 72, 74.

19. *See Hardwick*, 478 U.S. at 194-96.

20. *Miller*, 647 F.2d at 83 (Norris, J., dissenting) [emphasis added].

21. *See infra* chap. 7, note 19 (opinion polls show over 79% of Americans disapprove of homosexual relations).

22. *See, e.g., Dronenburg*, 741 F.2d 1388; *Rich*, 735 F.2d 1220; *Matlovich*, 591 F.2d 852.

23. *But see supra* chap. 3, notes 9 to 13 and accompanying text.

24. *See, e.g., Ben-Shalom*, 881 F.2d at 464; *Rich*, 735 F.2d at 1227-28, n.7; *Woodward*, 871 F.2d at 1076; *Pruitt*, 659 F. Supp. at 627; *Berg*, 591 F.2d 849; *Johnson*, 617 F. Supp. 170.

25. United States Information Agency v. Krc, 905 F.2d 389 (D.C. Cir. 1990).

26. Doe v. Casey, 796 F.2d 1508, 1524 (D.C. Cir. 1986)

27. *See, e.g., Alberico*, 783 F.2d at 102.

28. *Ben-Shalom I*, 489 F. Supp. at 971.

29. *Lindenau*, 663 F.2d at 71.

30. Cortright v. Resor, 447 F.2d 245, 251 (2d Cir. 1971), *cert. denied*, 405 U.S. 965 (1972).

31. *See Roth*, 408 U.S. at 577.

32. *Id.*

33. *See supra* chap. 2, notes 3 to 6.

34. *Gilliard*, 483 U.S. at 604 (citing Bowen v. Public Agencies Opposed to Social Security Entrapment, 477 U.S. 41 (1986) [emphasis added].

35. *Id.*

36. *See, e.g., Rich*, 735 F.2d 1220; *cf.* Ybarra v. Bastian, 647 F.2d 891, 893 (9th Cir. 1981) (property interests are extinguished when employee is disqualified for future employment) (citing *Beller*, 632 F.2d at 805).

37. Doe v. Garrett, 903 F.2d 1455 (11th Cir. 1990) (plaintiff challenged his exclusion from military service based on his HIV infection).

38. Saal v. Middendorf, 427 F. Supp. 192, 198 (N.D. Cal. 1977), *reversed sub nom Beller*, 632 F.2d 788.

39. *E.g., Rich*, 735 F.2d at 1227 ("plaintiff himself publicized his homosexuality and the circumstances of his discharge").

40. *Beller*, 632 F.2d at 806.

41. Neal v. Sec'y of the Navy , 472 F. Supp. 763 (E.D. Pa. 1979).

42. *Id.; Beller*, 632 F.2d at 806 ("mere fact of discharge from a government position does not deprive a person of a liberty interest").

43. *Id.*

44. *But cf. Matlovich*, 591 F.2d 852 (finding Air Force had not adequately followed procedures for implementing homosexual exclusion policy).
45. Davis, *supra* chap. 2, note 28, at 91.
46. Cleburne v. Cleburne Living Center, Inc., 473 U.S. 432, 440-41 (1985).
47. *High Tech Gays*, 895 F.2d at 573.
48. *Gilliard*, 483 U.S. at 602-03 (citing Lyng v. Castillo, 477 U.S. 635, 638 (1986) (classes of individuals excluded from the Federal Food Stamp Program were not suspect classes).
49. *E.g., Ben-Shalom*, 881 F.2d at 465.
50. Phillips, *Gay Rights Bill Goes Before House Today*, Boston Globe, Mar. 27, 1989, at 49.
51. *See, e.g., Impact of City Sex-Bias Laws Tough to Gauge*, Los Angeles Times, Oct. 30, 1989, at B1, col. 2.
52. Paid Political Advertisement, Take Back Tampa Campaign Committee, Joseph McAuliffe, Treasurer.
53. Wall Street Journal, July 18, 1991, at B1 (Simmons Market Research Bureau, U.S. Census data. Figures are from 1988, latest years available at time of article).
54. *Id.*
55. *Id.*
56. *Id.*
57. Take Back Tampa Campaign, *supra* chap. 3, note 52.
58. *Padula*, 822 F.2d at 103; *accord Baker*, 769 F.2d at 292; *Rich*, 735 F.2d at 1229; *Ben-Shalom*, 881 F.2d at 465-66.
59. *Id.* at 465.
60. *Id.* at 466.
61. *Padula*, 822 F.2d at 102 (quoting *San Antonio School Dist.*, 411 U.S. at 93).
62. *See* De Santis v. Pacific Telephone and Telegraph Co., 608 F.2d 327, 333 (9th Cir. 1979) [emphasis added].
63. *Bowen*, 483 U.S. at 602 (citing *Massachusetts Bd. of Retirement*, 427 U.S. at 313-14; *but see, e.g., High Tech Gays*, 909 F.2d at 377 (Canby, J., dissenting) (Supreme Court has stated the standard for suspect classification variously).
64. *See, e.g.*, Davis, *supra* chap. 2, note 28, at 93.
65. *Id.* at 57-63.
66. Schweiker v. Wilson, 450 U.S. 221, 229 (1981) [emphasis added].
67. *See infra* chap. 4, notes 43 to 46 and 55 to 62 (discussing *ex post* and *ex ante* analyses) and accompanying text (irrelevance of "fault" in military personnel policies).

68. *See, e.g.*, *High Tech Gays*, 909 F.2d at 377 (Canby, J., dissenting).

69. *Cf. Frontiero*, 411 U.S. 677 (gender not a suspect classification); *Trerice v. Pedersen*, 769 F.2d 1398, 1402-03 (9th Cir. 1985) (rejecting claim military prisoners were a protected class on the basis prisoners did not possess "discrete, insular and immutable characteristics comparable to those characterizing classes such as race, national origin and sex").

70. *High Tech Gays*, 895 F.2d at 573; *Woodward*, 871 F.2d at 1076; *Padula*, 822 F.2d at 103.

71. *Ben-Shalom*, 881 F.2d at 463 (discussing the district court's analysis).

72. *High Tech Gays*, 909 F.2d at 377 (Canby, J., dissenting).

73. *Cf. Khalsa*, 759 F.2d 1411 (Sikh excluded because religious precepts precluded meeting military appearance standards); *Goldman*, 475 U.S. 503 (Orthodox Jew excluded for refusing to abide by uniform standards); *see also* O'Lone v. Estate of Shabazz, 482 U.S. 341 (1987) (upholding prison regulations preventing attendance at religious service).

74. *See Dicicco*, 873 F.2d 910 (plaintiff initially excluded from military because he did not speak English).

75. *See, e.g., Smith*, 763 F.2d 1322 (plaintiff excluded from the military because he was missing a finger).

76. *See Ben-Shalom*, 881 F.2d at 457 (plaintiff claimed homosexuals were a discrete and insular group for equal protection purposes; the court focused on the group as defined by homosexual behavior).

77. *Norton*, 417 F.2d at 1167, n.28.

78. *Cyr*, 439 F. Supp. at 703.

79. *Id.* at 704; M. Humphrey, *supra* chap. 2, note 3, at *ix* ("[g]ays cannot, after all, be detected simply by the color of their skin").

80. Remafedi, *Adolescent Homosexuality: Psychosocial and Medical Implications*, 79 Pediatrics 331-37 (1987).

81. Smith v. Liberty Mutual Ins. Co., 569 F.2d 325 (5th Cir. 1978).

82. Davis, *supra* chap. 2, note 28, at 64.

83. *Nat'l Gay Task Force*, 729 F.2d 1270 (citing *Frontiero*, 411 U.S. 677).

84. *Id.*

85. *E.g., Rich*, 735 F.2d at 1229; *Ben-Shalom*, 881 F.2d at 464; *Hatheway*, 641 F.2d at 1382; *Woodward*, 871 F.2d at 1076; *Padula*, 822 F.2d at 103; *Nat'l Gay Task Force*, 729 F.2d at 1273; *DeSantis*, 608 F.2d at 333; *High Tech Gays*, 895 F.2d at 571.

86. *See Lindenau*, 663 F.2d 68; *cf. Gilliard*, 483 U.S. 587 (families are not a suspect class).

87. Holoway v. Arthur Anderson & Co., 577 F.2d 659 (9th Cir. 1977); *accord* Doe v. Alexander, 510 F. Supp. 900.
88. Plyler v. Doe, 457 U.S. 202, 223-34 (1982) (undocumented aliens were not a suspect class, but statute did not meet the rational basis test).
89. *Trerice*, 769 F.2d 1398.
90. *Schweiker*, 450 U.S. 221.
91. *Stanglin*, 490 U.S. at 25.
92. Massachusetts Bd. of Retirement v. Murgia, 427 U.S. 307 (1976). Also see Papasan v. Allain, 478 U.S. 265 (1986) (the wealthy are not a suspect class).
93. *E.g.*, *Lindenau*, 663 F.2d 68 (single parents excluded from military accession).
94. *Schweiker*, 450 U.S. 221; Weinberger v. Wiesenfeld, 420 U.S. 636, 638, n.2 (1975); *see Beller*, 632 F.2d at 810; *Ben-Shalom*, 881 F.2d at 463-64; *Woodward*, 871 F.2d at 1075, n.7.
95. *Ben-Shalom*, 881 F.2d at 464; *accord Padula*, 822 F.2d at 97; *High Tech Gays*, 895 F.2d at 570-71.
96. *High Tech Gays*, 909 F.2d 375 (Canby, J., dissenting).
97. Doe v. Commonwealth's Attorney, 425 U.S. 901 (1976).
98. Davis, *supra* chap. 2, note 28, at 91.
99. *Ben-Shalom*, 881 F.2d at 464; *accord Padula*, 822 F.2d at 103; *Woodward*, 871 F.2d at 1076.
100. Davis, *supra* chap. 2, note 28, at 91.
101. Olmstead v. United States, 277 U.S. 438, 478 (1928) (Brandeis, J., dissenting).
102. Davis, *supra* chap. 2, note 28, at 91.
103. *Ben-Shalom I*, 489 F. Supp. at 975.
104. *O'Lone*, 482 U.S. 342; Ogden v. United States, 758 F.2d 1168 (7th Cir. 1985).
105. *Rich*, 735 F.2d at 1228.
106. *Beller*, 632 F.2d at 810.
107. *Miller*, 647 F.2d at 83 (Norris, J., dissenting).
108. Homosexuality is only one in a whole spectrum of possible sexual identities that might be established as at the core of one's personality, self-image, or identity. *See infra* chap. 4, notes 140 to 143 and accompanying text.
109. *Gilliard*, 483 U.S. at 604-05.
110. 475 U.S. at 507-08.
111. *See Khalsa*, 759 F.2d 1411 (Sikh excluded from the military).

112. Davis, *supra* chap. 2, note 28, at 103.
113. *Dillard*, 652 F.2d 316 (plaintiff would have been excluded from military service had she obeyed ruled to disclose she was a single parent).
114. Manual for Courts-Martial, 1984, Part IV, para. 14c(2)(a)(iv).
115. *See infra* chap. 5, notes 12 to 17 and accompanying text.
116. *E.g.*, *Ben-Shalom*, 881 F.2d at 464; *Woodward*, 871 F.2d at 1075; *Dronenburg*, 741 F.2d at 1398.

Chapter Four

1. *Voorhees*, 16 C.M.R. at 107.
2. *See Lindenau*, 663 F.2d at 73.
3. *Ben-Shalom*, 881 F.2d at 461 (quoting *Goldman*, 475 U.S. at 507); *but cf. Miller*, 647 F.2d 80 (Norris, J., dissenting) (suggesting a due process analysis focused on the significance and intimacy of the personal decision to the individual).
4. McGinnis v. Royster, 410 U.S. 263, 270 (1973).
5. *Cleburne*, 473 U.S. at 440.
6. *Stanglin*, 490 U.S. at 26.
7. *Cleburne*, 473 U.S. at 440.
8. *Stanglin*, 490 U.S. at 27.
9. *Id.* While *Stanglin* addressed a state policy, the same analysis applies to federal policies: the courts should not substitute their view of wise social—or military—policy for the view of the Congress or the executive.
10. *Ben-Shalom*, 881 F.2d at 464-65; *Woodward*, 871 F.2d at 1076; *Rich*, 735 F.2d at 1229; *Dronenburg*, 741 F.2d at 1398; *accord Padula*, 882 F.2d at 103; *High Tech Gays*, 895 F.2d at 574; *Baker*, 769 F.2d at 292.
11. Vuono, *Training and the Army of the 1990s*, 1 Mil. Rev. 2-9 (1991) at 4.
12. *See supra* chap. 3, note 29 and accompanying text.
13. *High Tech Gays*, 895 F.2d at 576.
14. *Cleburne*, 473 U.S. at 440. *Espinoza*, 814 F.2d at 1097-98.
15. *Espinoza*, 814 F.2d at 1097-98.
16. *Id.* at 1097 [emphasis added].
17. *Dronenburg*, 741 F.2d at 1398; *cf. Gilliard*, 483 U.S. at 587, 599-601, n.5 (common sense and experience is used to formulate statutory schemes); *High Tech Gays*, 895 F.2d at 566 (the ultimate decision on security clearances "must be an overall common sense determination based upon all available facts").

18. *Id.* at 1398.
19. *Aumiller*, 434 F. Supp. at 1290.
20. *Baker*, 553 F. Supp. at 1129-31.
21. *Id.* at 1129, n.2.
22. *See* Maier v. Orr, 754 F.2d 973 (Fed. Cir. 1985).
23. *See* United States v. Benedict, 27 M.J. 253 (C.M.A. 1988).
24. *Cf. Mathews*, 424 U.S. 319 (even where process is constitutionally due, fiscal and administrative burdens are factors in determining how much process is "enough").
25. *See, e.g., Rich*, 735 F.2d at 1227-28, n.7.
26. *See, e.g., Matthews*, No. 82-0216-P, slip op. (D. Me. Apr. 3, 1984) (testimony of Major General H. Norman Schwarzkopf and declaration of Major General Kenneth L. Peek, in their capacities as former Director for Military Personnel Management and Assistant Deputy Chief of Staff for Personnel for the Army and Director of Personnel Plans, Office of the Deputy Chief of Staff, Manpower and Personnel, Department of the Air Force, respectively), *remanded on other grounds*, 755 F.2d 182 (1st Cir. 1985); *Beller*, 632 F.2d 788 (affidavit of Assistant Chief of Naval Personnel); *cf. Stanglin*, 490 U.S. 19 (rational basis demonstrated by testimony of an urban planner and a city police officer); *Espinoza*, 814 F.2d at 1097-98 (rational basis demonstrated by testimony of prison officials).
27. This "platitude" language was widely used by plaintiffs and their counsel. *See Glines*, 444 U.S. at 368-69 (Brennan, J., dissenting) (rejecting argument military had a substantial interest in good order, discipline, and morale).
28. *E.g., Beller*, 632 F.2d at 811; *Ben-Shalom*, 881 F.2d at 465.
29. *Id.* at 460.
30. *Beller*, 632 F.2d at 811; *accord Dronenburg*, 741 F.2d at 1398; *Ben-Shalom*, 881 F.2d at 465; *Woodward*, 871 F.2d at 1076; *Pruitt*, 659 F. Supp. at 527; *Rich*, 735 F.2d at 1227-28, n.7; *Hatheway*, 641 F.2d at 1382 ("those who engage in homosexual acts severely compromise the government's ability to maintain [a strong military] force").
31. *Cf. High Tech Gays*, 895 F.2d at 574 (citing Celotex Corp. v. Catrett, 477 U.S. 317, 331 (1986) (Brennan, J., dissenting)); *Harper*, 877 F.2d 728. In *Harper*, the court explained,

> The moving party has the burden of demonstrating the absence of a genuine issue of fact for trial. . . . If the moving party satisfied this burden, the opponent must set forth specific facts showing that there remains a genuine issue for trial. . . . However, no

defense to an insufficient showing is required. . . . [A]n issue of fact is only a genuine issue if it can reasonably be resolved in favor of either party.

Id. Most challenges to the homosexual exclusion policy were decided on the Army's motion to dismiss, or in the alternative, for summary judgment.

32. *See, e.g., Rich,* 735 F.2d at 1227-28, n.7.
33. *See, e.g., Matlovich,* 591 F.2d 852.
34. *Beller,* 632 F.2d at 81.
35. "A non-moving party who bears the burden of proof at trial to an element essential to its case must make a showing sufficient to establish a genuine dispute of fact with respect to the existence of that element of the case or be subject to summary judgment." *Harper,* 877 F.2d at 731 (citing *Celotex,* 477 U.S. at 322); *accord High Tech Gays,* 895 F.2d at 574.

Further, the "burden to demonstrate a genuine issue of fact increases where the *factual context makes the non-moving party's claim implausible."* *Harper,* 877 F.2d at 731 (citing Matshushita Elec. Indus. Co., Ltd. v. Zenith Radio Corp., 475 U.S. 574, 587 (1986) [emphasis added]).

Plaintiffs generally did not attempt to prove—or show a genuine dispute of fact as to whether—homosexuality is incompatible with military service. Most plaintiffs had admitted to homosexual acts, and no plaintiff made a forthright claim of celibacy. *See, e.g., Woodward,* 871 F.2d at 1074, n.6; *Steffan,* 733 F. Supp. at 123. Since the detrimental effect of homosexual conduct on the military was conceded, *see, e.g., Matthews,* 755 F.2d at 183, even if a status-conduct dichotomy was sensible and accepted, the factual context made plaintiffs' claims implausible.

36. Although this analysis is implicit in all litigation, few courts expressly analyzed the positions and relative burdens of the moving and non-moving parties, or whether evidence had been presented. The *High Tech Gays* court was an exception. There, plaintiffs challenged a Department of Defense (DoD) policy on investigating homosexuals who applied for security clearances. Plaintiffs won summary judgment below. The appellate court, however, held summary judgment had been improperly granted because plaintiffs'

affidavits and evidence fail to make a sufficient showing that the DoD *does not have a rational basis for its* [*policy*] or that there is a

> *genuine issue of material fact for trial.* . . . Nor have plaintiffs offered
> *"any significant probative evidence tending to support their complaint."*

Id. at 575 (citations omitted) [emphasis added].

37. *Miller,* 647 F.2d at 85 (Norris, J., dissenting).
38. *Ben-Shalom,* 881 F.2d at 462 (citing *O'Brien,* 391 U.W. 367).
39. *Toth v.* Quarles, 350 U.S. 11, 17 (1955).
40. Noyd v. McNamara, 378 F.2d 538, 540 (10th Cir. 1967).
41. *Mathews,* 424 U.S. at 332.
42. *Beller,* 632 F.2d at 812 [emphasis added].
43. *See* Easterbrook, *The Supreme Court, 1983 Term—Foreword: The Court and the Economic System,* 98 Harv. L. Rev. 4 (1984).
44. *Id.* Even if plaintiffs' arguments were responsive to policy concerns, arguments are not evidence and evidence is required to negate the Army's showing of a rational basis for the determination homosexuality is incompatible with military service. See *supra* chap. 4, notes 31 to 36 and accompanying text.
45. *Clark v. Community for Creative Non-Violence (CCNV),* 468 U.S. 288 (1984).
46. Easterbrook, *supra* chap. 4, note 43 [emphasis added].
47. *See, e.g.,* Davis, *supra* chap. 2, note 28, at 58.
48. *High Tech Gays,* 909 F.2d at 377 (Canby, J., dissenting).
49. *See, e.g., Health Care Needs of a Homosexual Population,* 248 J.A.M.A. 736-39 (1982).
50. *Rich,* 735 F.2d at 1224-25, n.1.
51. *Alberico,* 783 F.2d at 1028.
52. *Rich,* 735 F.2d at 1228.
53. *Dronenburg,* 741 F.2d at 1391.
54. *Gilliard,* 483 U.S. at 601.
55. *See, e.g.,* Davis, *supra* chap. 2, note 28, at 57-63; *Baker,* 553 F. Supp. at 1121, *rev'd,* 769 F.2d 289.
56. *Id.* at 1129.
57. *Aumiller,* 434 F. Supp. at 1290.
58. *Schweiker,* 450 U.S. at 229, n.11.
59. Davis, *supra* chap. 2, note 28 at 58.
60. *High Tech Gays,* 909 F.2d at 377 (Canby, J., dissenting).
61. Friedland, *AIDS and Compassion,* 259 J.A.M.A. 1898-99 (1988).
62. *Cf. High Tech Gays,* 895 F.2d at 575.
63. *Robinson,* 370 U.S. at 667, n.9.
64. *Hardwick,* 478 U.S. at 197 (Burger, C.J., concurring). The Chief Justice

went on to find "nothing in the Constitution depriving a State of the power to enact the statute [proscribing sodomy]." *Id.*

65. *Ben-Shalom*, 881 F.2d at 462.

66. *Toth*, 350 U.S. at 17.

67. *Rich*, 735 F.2d at 1228, n.7; *but cf. Matlovich*, 591 F.2d 852.

68. *Schweiker*, 450 U.S. at 235.

69. *See supra* chap. 2, note 3 and chap. 3, notes 32 to 35 and accompanying text.

70. This is a paraphrase of Justice Harlan's comments in In re Winship, 397 U.S. 358, 372 (concurring opinion) ("it is far worse to convict an innocent man than to let a guilty man go free").

71. *Gilliard*, 483 U.S. at 500 [emphasis added].

72. *Beller*, 632 F.2d at 808, n.20 (citing Weinberger v. Salfi, 422 U.S. 749 (1975)).

73. *Schweiker*, 450 U.S. at 234 (quoting *Massachusetts Bd. of Retirement*, 427 U.S. at 314).

74. *Cf. Mathews*, 424 U.S. 319.

75. *Beller*, 632 F.2d at 809, n.20. *Id.* at 808, n.20.

76. *Id.* at 808, n.20.

77. *Id.*

78. Adams v. Laird, 420 F.2d 230, 239 (D.C. Cir. 1969), *cert. denied*, 397 U.S. 1039 (1970).

79. *See, e.g., Ben-Shalom*, 881 F.2d at 464.

80. *Steffan*, 733 F. Supp. 121.

81. Goldsmith, *HIV Prevalence Data Mount, Patterns Seen Emerging by End of This Year*, 260 J.A.M.A. 1829-30 (1988).

82. M. Humphrey, *supra* chap. 2, note 3, at 111.

83. *Id.* at 11.

84. *Id.* at *xiii*, 4, 9, 12, 31, 33, 37, 45, 48, 74, 85, 91, 98, 105, 111, 125, 129, 136, 141, 145, 153, 164, 169, 176, 181, 202, 225, 229, 239, 245, and 251.

85. *See, e.g., Berg*, 591 F.2d 849; *Matlovich*, 591 F.2d 852.

86. *Id.* at 851.

87. *Id.*

88. Doe v. Casey, 796 F.2d at 1523.

89. *Ben-Shalom*, 881 F.2d at 464.

90. *Beller*, 632 F.2d at 809, n.20.

91. *Ben-Shalom*, 881 F.2d at 464.

92. M. Humphrey, *supra* chap. 2, note 3, at 111 [emphasis added].

93. *Id.* at 45 [emphasis added].

94. *Id.* at 229 [emphasis added].

95. M. Humphrey, *supra* chap. 2, note 3, at 108-18.

96. *Hatheway*, 641 F.2d at 1378; *cf. also Neal*, 472 F. Supp. 763.

97. AR 27-10, para. 6-5 (Military Justice).

98. *Salfi*, 422 U.S. at 776.

99. *Id.*

100. *See supra* chap. 2, note 68.

101. *Parker*, 417 U.S. at 765.

102. *Dronenburg*, 741 F.2d at 1397.

103. *Miller*, 647 F.2d at 86 (Norris, J., dissenting) [emphasis added].

104. *Hardwick*, 478 U.S. at 195. The Court declined to disregard morality and stated they were "unpersuaded that the sodomy laws of some 25 states should be invalidated on this basis." *Id.*

105. *Dronenburg*, 741 F.2d at 1307.

106. *Id.*

107. *Doe v. Commonwealth's Attorney*, 425 U.S. at 1202; *Hardwick*, 478 U.S. 186.

108. *Id.* at 195.

109. *Dronenburg*, 741 F.2d at 1392; see *infra* chap. 4, note 16 and accompanying text.

110. *E.g., Dronenburg*, 741 F.2d at 1397.

111. *Parker*, 417 U.S. at 765 (Blackmun, J. concurring) (citing Fletcher v. United States 26 Ct. Cl. 541, 562-63 (1891).

112. *See* United States v. Stockman, 17 M.J. 530, 532 (A.C.M.R. 1983) ("[c]onduct that deliberately violates an order issued to protect the status of Berlin by a soldier assigned to a unit specifically deployed to protect the status of Berlin is not merely 'bad', it is *dishonorable*. The appellant's approved sentence is fully warranted").

113. *See Glines*, 444 U.S. at 368-69 (Brennan, J., dissenting).

114. *Watkins*, 875 F.2d at 728 Ien banc) (Norris, J., concurring); *contra Ben-Shalom*, 881 F.2d at 465.

115. *High Tech Gays*, 895 F.2d at 578.

116. *Miller*, 647 F.2d at 88 (Norris, J., dissenting).

117. *Beller*, 632 F.2d at 81.

118. *Study: Integration of Dutch Navy Not Without Problems*, Stars and Stripes (European Edition), Nov. 30, 1992, at 4.

119. *Krc*, 905 F.2d at 939.

120. *Padula*, 822 F.2d at 104.

121. Sullivan v. Immigration and Naturalization Svc., 772 F.2d 609 (9th Cir. 1985).

122. *Ben-Shalom*, 881 F.2d at 464.

123. *Matthews*, No. 82-0216-P, slip op. (D. Me. Apr. 3, 1984).

124. *Saal*, 427 F. Supp. at 203, *rev'd sub nom Beller*, 632 F.2d 788 [emphasis added].

125. *See, e.g., Miller*, 647 F.2d 80 (Norris, J., dissenting from rejection of suggestion for rehearing en banc).

126. *Impact of City Sex-Bias Law Touch to Gauge*, Los Angeles Times, Oct. 30, 1989, at B1, col. 2 [emphasis added].

127. *Dronenburg*, 741 F.2d at 1397.

128. *Dronenburg*, 741 F.2d at 1397.

129. *Id.*

130. *Matthews*, No. 82-0216-P, slip op. (D.C. Me. Apr. 3, 1984).

131. *Commentary; Irvine's Human Rights Ordinance*, Los Angeles Times, Oct. 22, 1989, at B12, col. 2 [emphasis added].

132. Sheldon, *Acceptance of Homosexuals as "Minorities" is Common—but Wrong*, Los Angeles Times, Aug. 20, 1989, at 14, col. 1.

133. *A Waltz With Truth or Trouble?*, Los Angeles Times, Sep. 13, 1990, at E1, col. 4.

134. The Words of Martin Luther King, Jr., Newmarket Press, 95 (1983).

135. *See infra* chap. 5, notes 194 to 237 and accompanying text.

136. *See infra* chap. 5, notes 238 to 249 and chap. 6, notes 96 to 98 and accompanying text.

137. *See infra* chap. 6, note 137 and accompanying text.

138. *Id.*

139. *Dronenburg*, 741 F.2d at 1397.

140. *Dronenburg*, 747 F.2d at 1583.

141. *Id.*

142. *Rich*, 735 F.2d at 1228.

143. *See Dronenburg*, 741 F.2d at 1397, n.6; *cf. High Tech Gays*, 895 F.2d at 568.

144. *See, e.g., Minority Issues in AIDS*, 103 Pub. Health Rep. 91-4 (1988).

145. *Minority Issues in AIDS*, 103 Pub. Health Rep. 91-4 (1988) [emphasis added].

146. McCarthy, *The Role of the American Hospital Association in Combating AIDS*, 103 Pub. Health Rep. 273-77 (1988).

147. *Ulane*, 742 F.2d at 1084.
148. *Baker*, 553 F. Supp. at 1127, n.8.
149. *Cyr*, 439 F. Supp. at 699, n.2.
150. *See, e.g.*, Johnson v. San Jacinto Jr. College, 498 F. Supp. 555, 575 (S.D. Tex. 1980) and Wilson v. Swing, 463 F. Supp. 555 (M.D.N.C. 1978) (extramarital relationship not protected by constitutional right to privacy).
151. J.B.K., Inc. v. Caron, 600 F.2d 710 (8th Cir. 1979) (fundamental rights not violated by anti-prostitution statute).
152. *Padula*, 822 F.2d at 103.
153. *Dronenburg*, 741 F.2d at 1396.
154. The argument the homosexual exclusion policy illegitimately catered to private biases, *Watkins*, 875 F.2d at 728 (en banc) (Norris, J., concurring), cannot be limited to a challenge to the policy. If correct, it obviously must extend to the whole body of constitutional law regarding sexual behavior.
155. *Ben-Shalom*, 881 F.2d at 465 ("the concerns set forth in the [homosexual exclusion policy] cannot be so easily dismissed as mere prejudice"); *accord Dronenburg*, 741 F.2d 1388.
156. *See, e.g.*, *Beller*, 632 F.2d at 811.
157. Turner v. Safly, 482 U.S. 78, 89-90 (1987) [emphasis added].
158. *Cleburne*, 473 U.S. at 440 [emphasis added].
159. United States v. Matthews, 2 M.J. 881, 885 (A.C.M.R. 1976) (Costello, J., concurring).
160. *See, e.g.*, In re Grimley, 137 U.S. 147 (1890).
161. *Voorhees*, 16 C.M.R. at 107 (Latimer, J., concurring in part and dissenting in part).
162. J. McComsey & M. Edwards, The Soldier and the Law, Mil. Serv. Pub. Co., 5 & 7 (1941).
163. Air Force Manual 39-12, section 2-103(a).
164. *Matlovich*, 491 F.2d 852 [emphasis added].
165. *See, e.g.*, *Ben-Shalom I*, 489 F. Supp. 964; *Matthews*, No. 82-0216-P, slip op. (D. Me. Apr. 3, 1984).
166. 10 U.S.C. sec. 925.
167. *Ben-Shalom*, 881 F.2d at 460.
168. United States v. Crittenden, NMCM 84-2760, slip op. (N.M.C.M.R. Oct 5, 1984) (May, J., concurring) [emphasis added].
169. M.Humphrey, *supra* chap. 2, note 3, at 228 (interview with Starr, a former captain in the Air Force who was convicted for sodomy and a

false official statement at his trial by court-martial); *see* United States v. Scoby, 5 M.J. 160 (C.M.A. 1978); United States v. Lovejoy, 42 C.M.R. 210 (C.M.A. 1970) (convictions for sodomy).

Chapter Five

1. *Gilliard*, 483 U.S. at 601, n.15.
2. *Cf. Mathews*, 424 U.S. 319.
3. *Gilliard*, 483 U.S. at 599-601, nn.14 & 15.
4. *Id.* at 603 (quoting *Lyng*, 477 U.S. at 642) [emphasis added]; *see also* Brown v. Heckler, 589 F. Supp. 985 (E.D. Pa. 1984), *aff'd*, 760 F.2d 255 (3d Cir. 1985) (congressional presumption a stepparent will support his stepchildren was rational, although it was not a legal duty and even though not every stepparent did so).
5. United States v. Henderson, 36 C.M.R. 741 (A.F.B.R. 1965).
6. United States v. Williams, 8 M.J. 506, 509 (A.F.C.M.R. 1979) [footnotes omitted]; United States v. Martin, 5 C.M.R. 102 (C.M.A. 1952); United States v. Green, 22 M.J. 711 (A.C.M.R. 1986).
7. AR 635-200, para. 15-1 (Personnel Separations, Enlisted Personnel); also AR 635-100, para. 5 (Active Duty Officers); AR 135-175, para. 2 (Reserve Component Officers); AR 135-178, para. 10 (Reserve Component Enlisted Personnel).
8. *Voorhees*, 16 C.M.R. at 105-06 (Latimer, J., concurring part and dissenting in part).
9. *See* Chappell v. Wallace, 462 U.S. 296 (1983).
10. *See supra* chap. 2, note 60 and accompanying text.
11. *See* Ildefonso, *Oral Sex as a Risk Factor for Chlamydia-Negative Ureaplasma-Negative Nongonococcal Urethritis*, 15 Sex. Trans. Diseases 100-02 (1988).
12. Haverkos, *The Epidemiology of [AIDS] Among Heterosexuals*, 260 J.A.M.A. 1922-29 (1988).
13. Collier, *Cytomegalovirus Infection in Homosexual Men; Relationship to Sexual Practices, Antibody to Human Immunodeficiency Virus, and Cell-Mediated Immunity*, 82 Am. J. Med. 593-601 (1987).
14. Ross, *Illness Behavior Among Patients Attending a Sexually Transmitted Disease Clinic*, 14 Sex. Trans. Diseases 174-79 (1987).
15. Ostrow, *Sexually Transmitted Diseases and Homosexuality*, 10 Sex. Trans. Diseases 208-15 (1983).

16. Guinan, *Heterosexual and Homosexual Patients with the Acquired Immunodeficiency Syndrome*, 100 Annals Internal Med. 213-18 (1984).

17. Gold, *Unexplained Persistent Lymphadenopathy in Homosexual Men and the Acquired Immune Deficiency Syndrome*, 64 Med. 203-13 (1985).

18. Saltzman, *Reliability of Self-Reported Sexual Behavior Risk Factors for HIV Infection in Homosexual Men*, 102 Pub. Health Rep. 692-97 (1987).

19. Linn, *Recent Sexual Behaviors Among Homosexual Men Seeking Primary Medical Care*, 149 Archives Internal Med. 1685-90 (1989).

20. *See* Saltzman, *supra* chap. 5, note 18.

21. Goldsmith, *As AIDS Epidemic Approaches Second Decade, Report Examines What Has Been Learned*, 264 J.A.M.A. 431, 433 (1990).

22. Linn, *supra* chap. 5, note 19.

23. Hull, *Comparison of HIV-Antibody Prevalence in Patients Consenting to and Declining HIV-Antibody Testing in an STD Clinic*, 260 J.A.M.A. 935-38 (1988) [citations omitted].

24. Handsfield, *Trends in Gonorrhea in Homosexually Active Men*, 262 J.A.M.A. 2985-86 (1989).

25. Fleming, *Acquired Immunodeficiency Syndrome in Low-Incidence Areas; How Safe is Unsafe Sex?*, 258 J.A.M.A. 785-87 (1987).

26. *Young "Ignoring" AIDS Warning*, Daily Telegraph, Aug. 26, 1990, at 4.

27. *Relapses Into Risky Sex Found in AIDS Studies*, New York Times, June 22, 1990, at A18, col. 5.

28. *It's "The Pro-Sex '90's"; Gay Spirits Buoyant Again in San Francisco's Haunts*, Washington Times, June 22, 1990, at A1; *cf. infra* chap. 6, note 18 (homosexual leaders resist calls for safe sex, calling them premature and punitive).

29. *Young "Ignoring" AIDS Warning*, Daily Telegraph, Aug. 26, 1990, at 4.

30. *Relapses Into Risky Sex Found in AIDS Studies*, New York Times, June 22, 1990, at A18, col. 5.

31. *Safe Sex Relapse Poses AIDS Threat*, UPI, June 21, 1990.

32. *See* Hellinger, *Updated Forecasts of the Costs of Medical Care for Persons with AIDS*, 105 Pub. Health Rep. 1-12 (1990) ("back calculating" size of HIV infection pool).

33. Haverkos, *supra* chap. 5, note 12.

34. *Relapses Into Risky Sex Found in AIDS Studies*, New York Times, June 22, 1990, at A18, col. 5.

35. Aral, *Demographic Effects on Sexually Transmitted Diseases in the 1970s: The Problem Could be Worse*, 10 Sex. Trans. Diseases 1000-01 (1983).

36. Remafedi, *supra* chap. 3, note 80.

37. *Book Directs Homosexuals to Local College "Cruising" Spots,* Washington Times, Feb. 23, 1990, at A1. *See Campus Life, infra* chap. 7, note 10 and accompanying text.

38. M. Humphrey, *supra* chap. 2, note 3, at 7, 29, 52, 57, 65-66, 68, 88, 89-90, 119-20, 201, & 228.

39. *Id.* at 57.

40. *Id.* at 228.

41. *Id.* at 201.

42. *Id.* at 90.

43. Biggar, *Low T-Lymphocyte Ratios in Homosexual Men,* 251 J.A.M.A. 1441-46 (1984).

44. *Partner Notification for Preventing [HIV] Infection—Colorado, Idaho, South Carolina, Virginia,* 260 J.A.M.A. 661315 (1988).

45. Remafedi, *supra* chap. 3, note 80.

46. *Health Care Needs of a Homosexual Population,* 248 J.A.M.A. 736-39 (1982).

47. *See* Gold, *supra* chap. 5, note 17.

48. *Id.*

49. Francis, *The Prevention of [AIDS] in the United States,* 257 J.A.M.A. 1357-66 (1987).

50. *Health Care Needs, supra* chap. 5, note 46.

51. Ross, *supra* chap. 5, note 14.

52. *High Tech Gays,* 895 F.2d at 580.

53. Ernst, *Characteristics of Gay Persons with Sexually Transmitted Disease,* 12 Sex. Trans. Diseases 59-63 (1985).

54. *Id.*

55. *See, e.g.,* Ramstedt, *Contact Tracing for Human Immunodeficiency Virus (HIV) Infection,* 17 Sex. Trans. Diseases 37-41 (1990).

56. Rutherford, *Contact Tracing and the Control of Human Immunodeficiency Virus Infection,* 259 J.A.M.A. 3609-10 (1988).

57. Guinan, *supra* chap. 5, note 16.

58. Ildefonso, *supra* chap. 5, note 11.

59. Walters, *Sexual Transmission of Hepatitis A in Lesbians,* 256 J.A.M.A. 594 (1986).

60. St. Lawrence, *Differences in Gay Men's AIDS Risk Knowledge and Behavior Patterns in High and Low AIDS Prevalence Cities,* 104 Pub. Health Rep. 391-95 (1989).

61. Stamm, *The Association Between Genital Ulcer Disease and Acquisition of HIV Infection in Homosexual Men,* 206 J.A.M.A. 1429-33 (1988).

62. St. Lawrence, *supra* chap. 5, note 60.

63. Walters, *supra* chap. 5, note 59 (citing Corey, *Sexual Transmission of Hepatitis A in Homosexual Men: Incidence and Mechanism*, 302 New Eng. J. of Med. 435-38 (1980).

64. Goilav, *Vaccination Against Hepatitis B in Homosexual Men*, 87 Am. J. Med. 3A-21S-3A-25S (1989); *accord* Kingsley, *Sexual Transmission Efficiency of Hepatitis B Virus and Human Immunodeficiency Virus Among Homosexual Men*, 264 J.A.M.A. 230-34 (1990).

65. Walters, *supra* chap. 5, note 59.

66. Roffman, *Continuing Unsafe Sex: Assessing the Need for AIDS Prevention Counseling*, 105 Pub. Health Rep. 202-08 (1990).

67. Gottesman, *The Use of Water-soluble Contrast Enemas in the Diagnosis of Acute Lower Left Quadrant Peritonitis*, 27 Diseases Colon and Rectum 84-88 (1984); Kingsley, *Colorectal Foreign Bodies, Management Update*, 28 Diseases Colon and Rectum 941-44 (1985).

68. Barone, *Management of Foreign Bodies and Trauma of the Rectum*, 156 Surg. Gyn. Ob. 453-57 (1983).

69. *See* Linn, *supra* chap. 5, note 19.

70. Kingsley, *supra* chap. 5, note 67.

71. Barone, *supra* chap. 5, note 68.

72. *See* Linn, *supra* chap. 5, note 19.

73. Collier, *supra* chap. 5, note 13.

74. *Id.*

75. *See, e.g.,* Newell, *Volatile Nitrites; Use and Adverse Effects Related to the Current Epidemic of the Acquired Immune Deficiency Syndrome*, 78 Am. J. Med. 811-16 (1985).

76. *Id.*

77. *Health Care Needs, supra* chap. 5, note 46.

78. Ostrow, *supra* chap. 5, note 15.

79. Walters, *supra* chap. 5, note 59.

80. *Health Care Needs, supra* chap. 5, note 46.

81. Cornelis, *Factors Influencing the Risk of Infection with [HIV] in Homosexual Men, Denver 1982-1985*, 16 Sex. Trans. Diseases 95-102 (1989).

82. Goilav, *supra* chap. 5, note 64; *accord* Kingsley, *supra* chap. 5, note 64.

83. Barone, *supra* chap. 5, note 68.

84. Surawicz, *Intestinal Spirochetosis in Homosexual Men*, 82 Am. J. Med. 587-92 (1987).

85. Guinan, *supra* chap. 5, note 16.

86. Gold, *supra* chap. 5, note 17.

87. *Id.*
88. Ernst, *supra* chap. 5, note 53 [citations omitted].
89. Collier, *supra* chap. 5, note 13.
90. Alter, *Hepatitis B Virus Transmission Between Heterosexuals*, 256 J.A.M.A. 1307-10 (1986).
91. Ostrow, *supra* chap. 5, note 15.
92. Holly, *Anal Cancer Incidence: Genital Warts, Anal Fissure or Fistula, Hemorrhoids, and Smoking*, 81 J. Nat'l Cancer Inst. 1726-31 (1989).
93. Ross, *supra* chap. 5, note 14.
94. Hill, *HIV Infection Following Motor Vehicle Trauma in Central Africa*, 261 J.A.M.A. 3282-83 (1989).
95. *Id.*
96. *Cf.* Maier v. Orr, 754 F.2d 973 (Fed. Cir. 1985).
97. *Id.*
98. *Henderson*, 36 C.M.R. 741
99. Surawicz, *supra* chap. 5, note 84.
100. *Id.*
101. Kingsley, *supra* chap. 5, note 67.
102. *Id.* [citations omitted].
103. Gingold, *Simple In-Office Sphincterotomy with Partial Fissurectomy for Chronic Anal Fissure*, 165 Surg. Gyn. Ob. 46-8 (1987).
104. Barone, *supra* chap. 5, note 68.
105. Stiffman, *Behavioral Risks for [HIV] Infection in Adolescent Medical Patients*, 85 Pediatrics 303-10 (1990).
106. Holly, *supra* chap. 5, note 91.
107. *Id.; accord* Taxy, *Anal Cancer*, 113 Archives Path. Lab. Med. 1127-31 (1989).
108. Cobb, *Giant Malignant Tumors of the Anus*, 33 Diseases Colon and Rectum 135-38.
109. *Health Care Needs, supra* chap. 5, note 46.
110. Surawicz, *supra* chap. 5, note 84.
111. *Id.*
112. *Health Care Needs, supra* chap. 5, note 46.
113. Ostrow, *supra* chap. 5, note 15.
114. *Id.*
115. Margolis, *Sexually Transmitted Anal and Rectal Infections*, 161 Surg. Gyn. Ob. 41 (1985).
116. Surawicz, *supra* chap. 5, note 84.
117. Barone, *supra* chap. 5, note 68.

118. *Id.; see, e.g.,* Kingsley, *supra* chap. 4, note 39; Crass, *Colorectal Foreign Bodies and Perforation,* 142 Am. J. Surg. 85-8 (1981).
119. *Id.*
120. Barone, *supra* chap. 5, note 68.
121. Doe v. Alexander, 510 F. Supp. 900 (D. Minn. 1981).
122. *Health Care Needs, supra* chap. 5, note 46.
123. *Ostrow, supra* chap. 5, note 15.
124. *Id.*
125. Remafedi, *supra* chap. 3, note 80.
126. Raymond, *Addressing Homosexuals' Mental Health Problems,* 259 J.A.M.A. 19 (1988).
127. Barone, *supra* chap. 5, note 68; Ostrow, *supra* chap. 5, note 15.
128. Whipple v. Martinson, 256 U.S. 41, 45 (1921).
129. United States v. Wade, 15 M.J. 993 (N.M.C.M.R. 1983).
130. United States v. Miller, 16 M.J. 858 (N.M.C.M.R. 1983).
131. Bieber, *Book Reviews: T. O. Ziebod & J. E. Mongeon, Alcoholism and Homosexuality,* 171 J. of Nerv. and Mental Disease 755-56.
132. *Id.*
133. Ernst, *supra* chap. 5, note 53.
134. Lewis, *Drinking Patterns in Homosexual and Heterosexual Women,* 43 J. of Clinical Psych. 277-79 (1982).
135. Raymond, *supra* chap. 5, note 126.
136. *Id.*
137. Ernst, *supra* chap. 5, note 53.
138. *Id.* [emphasis added].
139. Bieber, *supra* chap. 5, note 131; *see supra* chap. 5, note 40 and accompanying text.
140. Rolfs, *Epidemiology of Primary and Secondary Syphilis,* 264 J.A.M.A. 1432-37 (1990).
141. Linn, *supra* chap. 5, note 19; *accord* Zylke, *Interest Heightens in Defining, Preventing AIDS in High-Risk Adolescent Population,* 262 J.A.M.A. 2197 (1989).
142. *Relapses Into Risky Sex Found in AIDS Studies,* New York Times, June 22, 1990, at A18, col. 5.
143. Remafedi, *supra* chap. 3, note 80.
144. Kingsley, *supra* chap. 5, note 64.
145. Gold, *supra* chap. 5, note 17; Kingsley, *supra* chap. 5, note 64.
146. Levine, *Retrovirus and Malignant Lymphoma in Homosexual Men,* 254 J.A.M.A. 1921-25 (1985).

147. Friedland, *Intravenous Drug Abusers and [AIDS]*, 145 Archives Internal Med. 1413-17 (1985).

148. Newell, *Volatile Nitrites; Use and Adverse Effects Related to the Current Epidemic of the Acquired Immune Deficiency Syndrome*, 78 Am. J. Med. 811-16 (1985).

149. M. Humphrey, *supra* chap. 2, note 3, at 41, 65, 85 & 88, 96-7, 103-06, 110, 137, 146, 159, 178 & 180.

150. *Id*. at 181.

151. *Id*. at 97.

152. *Id*. at 146.

153. *Secora*, 747 F. Supp. 406.

154. *Hatheway*, 641 F.2d 1376; *but see* M. Humphrey, *supra* chap. 2, note 3, at 111 (Hatheway's out-of-court account of his homosexual conduct while in the Army).

155. *High Tech Gays*, 895 F.2d at 569.

156. *See, e.g.*, United States v. Rasmussen, 4 M.J. 513 (C.G.C.M.R. 1976) (there was evidence the accused required constant supervision and lacked adaptability for military life).

157. United States v. Parker, 10 M.J. 849, 850 (N.C.M.R. 1981).

158. Raymond, *supra* chap. 5, note 126.

159. Linn, *supra* chap. 5, note 19.

160. Ross, *The Prevalence of Psychiatric Disorders in Patients with Alcohol and Other Drug Problems*, 45 Archives Gen. Psych. 1023-31 (1988).

161. Bieber, *supra* chap. 5, note 131.

162. Lewis, *Diagnostic Interactions; Alcoholism and Antisocial Personality*, 171 J. Nerv. Mental Diseases 105-13 (1983).

163. Raymond, *supra* chap. 5, note 126.

164. *Id*.

165. Remafedi, *supra* chap. 3, note 80.

166. *Baker*, 553 F. Supp. 1121.

167. *Rich*, 735 F.2d at 1223.

168. Raymond, *supra* chap. 5, note 126.

169. Atkinson, *Prevalence of Psychiatric Disorders Among Men Infected with [HIV]*, 45 Archives Gen. Psych. 859-64 (1988).

170. Flavin, *The [AIDS] and Suicidal Behavior in Alcohol-Dependent Homosexual Men*, 143 Am. J. of Psych. 1440-42 (1986).

171. *See, e.g.*, Bender, *Book Reviews: L.H. Silverman, Search for Oneness*, 172 J. Nerv. Mental Diseases 442-43 (1984).

172. *Dubbs*, 866 F.2d 1114.

173. *Rich*, 735 F.2d at 1225, n.4.
174. Raymond, *supra* chap. 5, note 126.
175. Remafedi, *supra* chap. 3, note 80; *see also Parents Fear Schools Teach Homosexuality*, Washington Times, Nov. 21, 1989, at A1 (citing the suicide rate among homosexual adolescents as two to six times higher than among heterosexual youth).
176. Raymond, *supra* chap. 5, note 126.
177. *Id.*
178. M. Humphrey, *supra* chap. 2, note 3, at 3, 30, 41, 50, 141, 146, 151-52, 159, 186.
179. *Baker*, 553 F. Supp. 1121.
180. United States v. Varraso, 21 M.J. 129 (C.M.A. 1985).
181. Handsfield, *supra* chap. 5, note 24.
182. Holland, *The Psychosocial and Neuropsychiatric Sequelae to the [AIDS] and Related Disorders*, 103 Annals Internal Med. 760-64 (1985) [emphasis added].
183. United States v. Dumford, 30 M.J. 137 (1990), *cert. denied*, 111 S. Ct. 150 (1991). *See* Wells-Petry, "Anatomy of an AIDS Case: Deadly Disease as an Aspect of Deadly Crime," *The Army Lawyer*, Dec. 1988.
184. Kegeles, *Intentions to Communicate Positive HIV-Antibody Status to Sex Partners*, 259 J.A.M.A. 216-17 (1988).
185. *Id.*
186. M. Humphrey, *supra* chap. 2, note 3, at 127.
187. *Id.* at 142-43.
188. *Id.* at 58.
189. *See infra* chap. 6, note 7 (describing isolation—especially from the homosexual community—after publicly disclosing homosexuality).
190. *Study: Integration of Dutch Navy Not Without Problems*, Stars and Stripes (European edition), Nov. 30, 1992 at 4.
191. Ross, *supra* chap. 5, note 14.
192. Doe v. Alexander, 510 F. Supp. 900, 905 (D. Minn. 1981) [emphasis added].
193. *Stanglin*, 490 U.S. at 27.
194. *Saal*, 427 F. Supp. at 201, n.11.
195. M. Humphrey, *supra* chap. 2, note 3, at *viii*.
196. *News Call*, Army, Feb. 1991, at 69.
197. *See* H. Summers, On Strategy: A Critical Analysis of the Viet Nam War, 205-06 (1984) ("security is essential to the preservation of combat power").

198. *High Tech Gays*, 895 F.2d at 577.

199. *Id.* at 576.

200. *Krc*, 905 F.2d at 389 [emphasis added].

201. *See, e.g.*, United States v. Hughes, 389 F.2d 535 (2d Cir. 1968) (homosexual conduct used by an extortion ring).

202. *High Tech Gays*, 895 F.2d at 563 [emphasis added].

203. M. Humphrey, *supra* chap. 2, note 3, at 217 [emphasis added].

204. *Above Suspicion*, The Times, Jan. 19, 1990, at Features.

205. *France; Out of Line*, Economist, Nov. 10, 1990, at 57 (Europe) & 61 (U.K.).

206. *The World is "Outing"; The New Gay Militants*, Washington Times, Sep. 13, 1990, at E1.

207. *Id.* [emphasis added].

208. *Dragging People Out of Their Closets, Is "Outing" a Search for Truth or New Kind of Media Witch Hunt?* Toronto Star, Aug. 25, 1990, at F3.

209. *Study: Integration of Dutch Navy Not Without Problems*, Stars and Stripes (European edition), Nov. 30, 1992 at 4.

210. *High Tech Gays*, 895 F.2d at 576.

211. *Id.*

212. *Dubbs*, 866 F.2d at 1116.

213. *Id.* [emphasis added].

214. Ashton v. Civiletti, 613 F.2d 923 (D.D.C. 1979).

215. *Dronenburg*, 741 F.2d at 1389.

216. Urban Jacksonville, Inc. v. Chalbeck, 765 F.2d 1085 (11th Cir. 1985).

217. *High Tech Gays*, 895 F.2d at 577 [original emphasis omitted and emphasis added].

218. *Dubbs*, 866 F.2d at 1116.

219. *Krc*, 905 F.2d at 393.

220. *High Tech Gays*, 895 F.2d at 575.

221. *Baker*, 553 F. Supp. at 1130.

222. *High Tech Gays*, 895 F.2d at 575.

223. M. Humphrey, *supra* chap. 2, note 3, at 116 [emphasis added].

224. McKeand v. Laird, 490 F.2d 1262 (9th Cir. 1973).

225. Holland, *supra* chap. 5, note 182.

226. Valdiviez v. United States, 884 F.2d 196 (5th Cir. 1989).

227. *High Tech Gays*, 895 F.2d at 577 [emphasis added].

228. *Id.*

229. *See, e.g., Dubbs*, 866 F.2d at 1116.

230. *Padula*, 822 F.2d at 104.

231. United States v. Benton, 7 M.J. 606, 608 (N.C.M.R. 1977).

232. *Voorhees*, 16 C.M.R. at 107.

233. *See, e.g.*, Webster v. Doe, 486 U.S. 592 (1988); *Dubbs*, 866 F.2d 1114; *Dorfmont*, 913 F.2d 1399; *Krc*, 905 F.2d 389; *McKeand*, 490 F.2d 1262; *Padula*, 822 F.2d 97; *High Tech Gays*, 895 F.2d 563; Doe v. Weinberger, 820 F.2d 1275 (D.C. Cir. 1987).

234. *Dorfmont*, 913 F.2d 1399 (Kozinski, J., concurring) (citing United States Navy v. Egan, 484 U.S. at 527).

235. *Dorfmont*, 913 F.2d at 1402.

236. Doe v. Weinberger, 820 F.2d at 1278.

237. Doe v. Casey, 796 F.2d at 1520 [emphasis added].

238. AR 635-200, para. 15-1.

239. United States v. Johnson, 4 M.J. 770 (A.C.M.R. 1978).

240. *Id.* at 771.

241. 489 U.S. 656 (1988).

242. *See, e.g.*, Nat'l Treas. Employees Union v. Yeutter, 918 F.2d 968 (D.C. Cir. 1990); Taylor v. O'Grady, 888 F.2d 1189 (7th Cir. 1989); Schaill v. Tippecanoe County School Corp., 864 F.2d 1309 (7th Cir. 1988); Nat'l Federation Fed. Employees v. Weinberger, 818 F.2d 935 (D.C. Cir. 1987); Unger v. Ziemniak, 27 M.J. 349 (C.M.A. 1989).

243. *See, e.g.*, Dufrin v. Spreen, 712 F.2d 1084, 1087 (6th Cir. 1983); Vaughan v. Ricketts, 859 F.2d 736, 747 (9th Cir. 1988); Torres v. Wis. Dept. of Health & Soc. Serv., 838 F.2d 944, 969 (7th Cir. 1988); McDonell v. Hunter, 809 F.2d 1302, 1329 (8th Cir. 1987).

244. *See, e.g.*, Rider v. Pennsylvania, 850 F.2d 982 (3d Cir. 1988) (hiring female guards to monitor female prisoners did not constitute unlawful discrimination against male guards because same-sex guards sometimes were required); Timm v. Gunter, 917 F.2d 1093 (8th Cir. 1990) (surveillance by opposite-sex guards was not unreasonable because intrusion on inmate privacy was minimal and guards had strong statutory rights to equal employment opportunities).

245. M. Humphrey, *supra* chap. 2, note 3, at 22.

246. *Id.* at 61.

247. *See* United States v. Johnson, 4 M.J. 770 (A.C.M.R. 1978) (voyeurism).

248. See *infra* notes 762 to 768 and accompanying text.

249. *See Timm*, 917 F.2d at 1093, n.13.

Chapter Six

1. *Matlovich*, 13 Fair. Empl. Prac. Cas. (BNA) 269 (1976).
2. M. Humphrey, *supra* chap. 2, note 3, at *x*.
3. *See* 10 U.S.C. sec. 3062(a).
4. Vuono, *supra* chap. 4, note 11, at 2-3.
5. Davis, *supra* chap. 2, note 28, at 104. Disclosure of homosexuality is already required in the military accession process. *But see* M. Humphrey, *supra* chap. 2, note 3. This disclosure requirement does not duplicate the situation proposed in the model, however, because presently homosexuality results in exclusion, whereas in the model homosexuality would not be a service-disqualifier.
6. Davis, *supra* chap. 2, note 28, at 104.
7. M. Humphrey, *supra* chap. 2, note 3, at 191-92.
8. *Dubbs*, 866 F.2d 1114.
9. Holland, *supra* chap. 5, note 182 [emphasis added].
10. *Cf. Harper*, 877 F.2d 728.
11. According to one published account, NAMBLA's motto is "SEX BY [AGE] EIGHT BEFORE IT'S TOO LATE." Pd. Pol. Ad., Take Back Tampa Campaign Committee, Nov. 1990.
12. *Harper*, 877 F.2d at 734.
13. Davis, *supra* chap. 2, note 28, at 104.
14. *See supra* chap. 4, note 84 and accompanying text.
15. *The Hands That Would Shape Our Souls*, Atlantic Monthly, Dec. 1990, at 79.
16. Holland, *supra* chap. 5, note 182.
17. *See Kramer vs. Kramer*, Los Angeles Times, June 20, 1990, at E1, col. 2.
18. *Id.*
19. *Baker*, 553 F. Supp. at 1130, *rev'd*, 769 F.2d 289.
20. *Relapses Into Risky Sex Found in AIDS Studies*, New York Times, June 22, 1990, at A18, col. 5.
21. *Young "Ignoring" AIDS Warning*, Daily Telegraph, Aug. 26, 1990, at 4.
22. *Hardwick*, 478 U.S. at 203 (Blackmun, J., dissenting).
23. *See, e.g., Now For a Drug Policy That Doesn't Do Harm*, New York Times, Dec. 18, 1990, at A24, col. 6 (advocating legalization of most narcotics).
24. Davis, *supra* chap. 2, note 28, at 106 and 108.

25. *Id.*
26. *High Tech Gays*, 909 F.2d at 382 (Canby, J., dissenting).
27. *See supra* chap. 4, note 84 (several former service members interviewed were involved in at least one investigation or administrative or judicial proceeding).
28. United States v. Van Slate, 14 M.J. 872, 876 (N.M.C.M.R. 1982) (May, J., concurring in part and dissenting in part).
29. *Beller*, 632 F.2d at 811.
30. McConnell v. Anderson, 451 F.2d 193, 196 (8th Cir. 1971).
31. 875 F.2d 699 (9th Cir. 1989).
32. *But see Ben-Shalom*, 881 F.2d at 465 (expressing serious doubts on the merits of the *Watkins* estoppel holding).
33. M. Humphrey, *supra* chap. 2, note 3, at *xiii*.
34. *Id.* at *xvii* & *xviii*.
35. M. Humphrey, *supra* chap. 2, note 3, at, *e.g.*, 186.
36. *Id.* at 11.
37. *Id.* at 15.
38. *Id.* at 18.
39. *Id.* at 86 & 87.
40. *Id.* at 229-30 & 235 [original emphasis].
41. United States v. Clayton, 38 C.M.R. 46 (C.M.A. 1967).
42. United States v. Eason, 21 M.J. 79 (C.M.A. 1985).
43. Davis, *supra* chap. 2, note 28, at 106.
44. 10 U.S.C. sec. 925.
45. *Id.*
46. Davis, *supra*, chap. 2, note 28, at 106. The "appearance or prospect of favoritism within a chain of command" seems to be encompassed by the element of "actual prejudice."
47. 483 U.S. 435 (1987).
48. *See supra* chap. 5, notes 38 to 42 and accompanying text.
49. *Id.*
50. Davis, *supra* chap. 2, note 28, at 104.
51. United States v. Ragan, 33 C.M.R. 331 (C.M.A. 1963).
52. United States v. Scott, 24 M.J. 578 (N.M.C.M.R. 1987).
53. Davis, *supra* chap. 2, note 28, at 104.
54. Solorio v. United States, 483 U.S. 435 (1987).
55. *Hardwick*, 478 U.S. at 195.
56. Davis, *supra* chap. 2, note 28, at 98.
57. *Id.* at 103.

58. *See, e.g.*, J. Reisman & E.W. Eichel, *Kinsey, Sex and Fraud: The Indoctrination of a People* (1991).

59. *Kinsey, Sex and Fraud*, Washington Times, Dec. 21, 1990, at B7.

60. *Id.; The Gay Science of Genes and Brains*, The Economist, Dec. 5, 1992 at 88 ("Researchers in the field doubt the oft-repeated figure that one in ten men is gay").

61. MacDonald, *High-Risk STD/HIV Behavior Among College Students*, 263 J.A.M.A. 3155-59 (1990).

62. *Contra* Davis, *supra* chap. 2, note 28, at 103.

63. *Baker*, 553 F. Supp. at 1129, n.14.

64. Davis, *supra* chap. 2, note 28, at 103.

65. *Id.* at 108.

66. *Id.*

67. *Dillard*, 652 F.2d at 318 & n.1 [emphasis added].

68. Davis, *supra* chap. 2, note 28, at 98 [emphasis added].

69. *Ben-Shalom*, 881 F.2d at 4607 (citing *Glines*, 444 U.S. at 360).

70. *Pruitt*, 659 F. Supp. at 627.

71. *See Dubbs*, 866 F.2d 1114.

72. *Gilliard*, 483 U.S. at 599.

73. Turner v. Safly, 482 U.S. 78 (1987).

74. *Id.* at 90-91 [emphasis added].

75. Guard and Charge of Quarters duties are the types of duties all soldiers, regardless of military occupational specialty, are required to perform unless excused from duty for a military reason. For example, Army legal clerks may pull guard duties and Army lawyers serve stints as the Staff Duty Officer. See United States v. Mayo, 12 M.J. 286 (C.M.A. 1982), for a description of duties and responsibilities of the Charge of Quarters.

76. *Ben-Shalom*, 881 F.2d at 465.

77. Foss, *Advent of the Nonlinear Battlefield: AirLand Battle-Future*, Army, Feb. 1991, at 21 [emphasis added].

78. AR 635-200, para. 15-1; Davis, *supra* chap. 2, note 28, at 101.

79. Davis, *supra* chap. 2, note 28, at 101.

80. *Woodward*, 871 F.2d at 1069-70.

81. *Miller*, 647 F.2d at 89 (Norris, J., dissenting) [emphasis added].

82. *Smith*, 569 F.2d 325.

83. 10 U.S.C. sec. 934.

84. *Cf., e.g., McConnell*, 451 F.2d 193.

85. *See Miller*, 647 F.2d at 89.

86. *Miller*, 647 F.2d at 89 (Norris, J., dissenting).
87. Davis, *supra* chap. 2, note 28, at 101.
88. *Id.* at 26 [citation omitted].
89. *Id.* at 23.
90. *Harper*, 877 F.2d at 730.
91. *Espinoza*, 814 F.2d at 1094.
92. *Beller*, 632 F.2d at 811-12; *Woodward*, 871 F.2d at 1069-70.
93. The model also is deficient to the extent the model contemplates "follower" training as well as leadership training."
94. AR 635-200, para. 15-1.
95. Davis, *supra* chap. 2, note 28, at 101-02.
96. *Id.* at 102.
97. *Id.*
98. *See generally id.*
99. *Desert Storm Had a Steamy Side, Poll Says*, Stars and Stripes (European edition), Oct. 3, 1992 at 1.
100. M. Watson, *Chief of Staff: Pre-War Plans and Preparations (US Army in World War II)* 12 (1950).
101. Chappell v. Wallace, 462 U.S. 296.
102. *See Ben-Shalom*, 881 F.2d at 460.
103. *See supra* chap. 4, note 165.
104. *Study: Integration of Dutch Navy Not Without Problems*, Stars and Stripes (European edition), Nov. 30, 1992 at 4.
105. *Cf. Acceptance of Homosexuals as "Minorities" is Common—but Wrong*, Los Angeles Times, Aug. 20, 1989, at 14, col. 1.
106. 632 F. Supp. 1319 (E.D. N.Y. 1986).
107. *Id.* at 1326 [emphasis added].
108. *Id.*
109. *Id.* [emphasis added].
110. Katcoff v. Marsh, 755 F.2d 223 (2d Cir. 1972) (detailing the history of the military chaplaincy from before the American Revolution).
111. *The Gay Tide of Catholic-Bashing*, U.S. News & World Report, Apr. 1, 1991, at 15.
112. *The Hands That Would Shape Our Souls*, Atlantic Monthly, Dec. 1990, at 80.
113. *See, e.g., Cyr*, 439 F. Supp. 697 (citing the fact the Metropolitan Community Church caters primarily to the spiritual and worldly problems of homosexuals).

114. *Methodists Rethink Beliefs on Gay Sex*, San Francisco Chronicle, Feb. 14, 1991, at A23.

115. *Von Hoffburg*, 615 F.2d 633.

116. *Lesbians Win Damages*, Stars and Stripes (European edition), Nov. 26, 1992 at 7.

117. Adams v. Howerton, 673 F.2d 1036, 1040-41 (9th Cir. 1982), *cert. denied*, 458 U.S. 1111 (1982).

118. *Sullivan*, 772 F.2d 609.

119. *Gay is Denied Spousal Right to Share in his Lover's Estate*, N.Y.L.J., Dec. 28,6 1990, at 1 (the court held persons of the same sex have no constitutional rights to enter into a marriage with each other).

120. *But see Pruitt*, 659 F. Supp. 625.

121. United States v. Johnson, 481 U.S. 681 (1987).

122. *A Waltz With Truth or Trouble?*, Los Angeles Times, Sep. 13, 1990, at E1, col. 4.

123. Williamson v. A. G. Edwards and Sons, Inc., 876 F.2d 69 (*th Cir. 1989).

124. United States v. Davis, 26 M.J. 445 (C.M.A. 1988) (male sailor court-martialed for wearing makeup and women's clothing).

125. *See supra* chap. 5, notes 233 to 237 and accompanying text.

126. Weston v. Lockheed Missiles & Space Co., 881 F.2d 814 (9th Cir. 1989).

127. *See supra* chap. 6, notes 31 to 32 and accompanying text.

128. *Gay Activists Protest HIV Testing in Military*, The Washington Times, Dec. 1, 1992, at A1.

129. *See supra* chap. 5, notes 28, 36, 41, 43, 49, 137 to 139 and accompanying text.

130. *See supra* chap. 6, note 7 and accompanying text.

131. *Rutgers Panel Seeks Way to Assure Gay Rights*, New York Times, Mar. 5, 1989, at 1, col. 2.

132. *Id*.

133. *Jones*, 632 F. Supp. 1319 (discussed *supra* chap. 6, notes 106 to 109 and accompanying text).

134. *The Hands That Would Shape Our Souls*, Atlantic Monthly, Dec. 1990, at 80 [emphasis added].

135. *Cyr*, 439 F. Supp. 697 (N.D. Tex. 1977) (citing *Mississippi Gay Alliance*, 536 F.2d at 1075-76 & n.4).

136. Gay Alliance v. Matthews, 544 F.2d 162 (4th Cir. 1976).

137. *Vuono, supra* chap. 4, note 11, at 4-5 [emphasis added].

138. *See supra* chap. 5, note 121 and accompanying text.
139. *Aumiller*, 434 F. Supp. at 1303, n.86
140. M. Humphrey, *supra* chap. 2, note 3, at 112.
141. *Voorhees*, 16 C.M.R. at 106.
142. *Goldman*, 475 U.S. at 450.
143. *Voorhees*, 16 C.M.R. at 108.
144. *Turner*, 482 U.S. 78; *see also O'Lone*, 482 U.S. 342.
145. *Turner*, 482 U.S. at 90.
146. AR 635-200, para. 15-1.

Chapter Seven

1. H. Summers, *supra* chap. 5, note 197, at 45.
2. *Id*. at 19 (foreword by Major General Jack N. Merritt).
3. *Id*. at 47.
4. *Voorhees*, 16 C.M.R. at 106.
5. *Cf., e.g.*, "McNamara's Project 100,000" (cited in *Dicicco*, 873 F.2d 910).
6. *See Dronenburg*, 741 F.2d at 1398.
7. *See infra* chap. 7, notes 19, 21 to 22 and accompanying text.
8. *See, e.g., Zablocki*, 434 U.S. at 399 (Powell, J., concurring in the judgment).
9. *The World is "Outing"; The New Gay Militants*, Washington Times, Sep. 13, 1990, at E1.
10. *Campus Life: Michigan; Gay Rights Group Protests Remarks by Official*, New York Times, at 41, col. 1.
11. Paid Political Advertisement, Take Back Campaign Committee, Joseph McAuliffe, Treasurer.
12. *Commentary; Irvine's Human Rights Ordinance*, Los Angeles Times, Oct 22, 1989, at B12, col. 2.
13. *Id*.
14. *Protesters Leave "Artworks" at Church Office*, UPI, Jul. 8, 1990.
15. *Impact of City Sex-Bias Laws Tough to Gauge*, Los Angeles Times, at B1, col. 2.
16. *Ulane*, 742 F.2d at n.11.
17. *Id*. at 1085; *accord DeSantis*, 608 F.2d 327.
18. *Kizas*, 707 F.2d 524, 542 (D.C. Cir. 1983).
19. *'60's Sexual Revolution Didn't Occur, Kinsey Study Says*, Los Angeles Times, June 30, 1989, at 18, col. 1 [emphasis added].

20. *A Requiem for Fashion*, Crain's New York Business, Apr. 9, 1990, at 1 [emphasis added].

21. *Less Support Seen for Gays in Military*, Stars and Stripes (European edition), Nov. 23, 1992 at 2.

22. *Homosexual Ban Wins Continual Support*, Stars and Stripes (European edition), Oct. 5, 1992 at 18.

23. *See supra* chap. 7, note 19.

24. *Rowland*, 730 F.2d at 444 (Edwards, J., dissenting).

25. *Judge Clears Way for Gay Night at Amusement Park*, Reuters North European Serv. (Los Angeles), Jul. 11, 1986.

26. *School Board Suspended Over Gay Issue*, Stars and Stripes (European edition), Dec. 3, 1992 at 4.

27. *Cf., e.g., supra* chap. 6, notes 116 to 119, 131 and chap. 7, note 8 and accompanying text.

28. *The Hands That Would Shape Our Souls*, Atlantic Monthly, Dec. 1990, at 80.

29. *Id.* at 81.

30. *Study: Integration of Dutch Navy Not Without Problems*, Stars and Stripes (European edition), Nov. 30, 1992 at 4.

31. H. Summers, *supra* chap. 5, note 197, at 36-7 (quoting Clausewitz, *On War*).

32. *Dronenburg*, 741 F.2d at 1396.

33. United States Constitution, Article I, Section 8.

Chapter Eight

1. *Orloff*, 345 U.S. at 93.

2. *See, e.g., Goldman*, 475 U.S. 503.

3. Gilligan v. Morgan, 413 U.S. 1, 10 (1973).

4. *Id.*

5. *Id.*

6. H. Summers, *supra* chap. 5, note 197, at 37 [emphasis added].

7. *Ben-Shalom*, 881 F.2d at 461 [emphasis added].

8. H. Summers, *supra* chap. 5, note 197, at 33 (quoting General Fred C. Weyand, Chief of Staff, U.S. Army, July 1976).

9. *Ben-Shalom*, 881 F.2d at 466 (homosexuals can seek a congressional determination on the military's homosexual exclusion policy, but the court will not substitute a mere judge-made rule for the Army's regulation).

10. *See* In re Grimley, 137 U.S. 147 (1890).
11. *See supra* chap. 7, notes 15 to 18 and accompanying text.
12. *See DeSantis*, 608 F.2d 327.
13. *Dronenburg*, 741 F.2d at 1396; *see supra* chap. 7, notes 15 to 26 and accompanying text.
14. *Matlovich*, 13 Fair Emp. Prac. Cas. (BNA) 269 (1976).
15. *See supra* chap. 5, note 48 and accompanying text.
16. *Dronenburg*, 741 F.2d at 1397 [emphasis added].
17. *Cruzan*, 110 S. Ct. 2841 (1990) (Scalia, J., concurring).
18. H. Summers, *supra* chap. 5, note 197, at 19.

Chapter Nine

1. Crocker, *supra* chap. 1, note 1, at *viii-ix*.
2. *See supra* chap. 2, notes 91 to 92 and accompanying text; *cf. supra* chap. 6, notes 81 to 82 and accompanying text (impact of flirting and so on).
3. *Saal*, 427 F. Supp. at 202, *rev'd sub nom Beller*, 632 F.2d 788.
4. *See supra* chap. 4, notes 43 to 46 and accompanying text.

Index

231